Thank you for choosing Basic Arithmetic for Adults!

We hope you will have a great time working through the exercises and improving your math skills, or just enjoying this book as a fun pastime! With its engaging and enjoyable exercises, this book is the perfect way to keep your brain sharp and your memory in top form. If you enjoyed this book, we encourage you to check out our other math-related titles. We appreciate your support and would be grateful if you could leave a review or feedback about the book. Your input will help us improve our products and better meet the needs of our readers.

TABLE OF CONTENTS

Rules

In each exercise, you will be required to perform six operations. Please follow the instructions from left to right, and do not apply the standard rules of priority of operations in mathematics. Your goal is to start with the number given and work your way through the tasks to reach the solution at the end. You can solve the problems in your head or use the free space provided to note intermediate results. Some of the exercises will be very easy, while others may be more challenging. Regardless of the difficulty, take your time and work through each problem carefully to ensure accuracy. A full solution key can be found at the end of the book if you need additional guidance or want to check your work.

Addition #1-50

#1	27	+ 26	+ 11	+ 70	+ 91	+ 81	+ 9	=	

#2	39	+ 61	+ 40	+ 91	+ 79	+ 23	+ 77	=	

#3	147	+ 41	+ 42	+ 45	+ 38	+ 3	+ 48	=	

#4	110	+ 84	+ 21	+ 41	+ 99	+ 46	+ 32	=	

#5	62	+ 88	+ 27	+ 53	+ 80	+ 82	+ 47	=	

#6	65	+ 54	+ 99	+ 31	+ 98	+ 14	+ 67	=	

#7	50	+ 41	+ 41	+ 65	+ 90	+ 22	+ 79	=	

#8	108	+ 98	+ 60	+ 25	+ 63	+ 36	+ 31	=	

#9	50	+ 9	+ 17	+ 80	+ 89	+ 15	+ 54	=	

#10	63	+ 68	+ 60	+ 14	+ 30	+ 61	+ 30	=	

#11	55	+ 33	+ 35	+ 82	+ 10	+ 74	+ 27	=	

#12	73	+ 87	+ 86	+ 70	+ 7	+ 21	+ 96	=	

| #13 | 71 | + 33 | + 74 | + 53 | + 90 | + 36 | + 23 | = | |

| #14 | 138 | + 25 | + 98 | + 18 | + 49 | + 29 | + 72 | = | |

| #15 | 98 | + 66 | + 56 | + 3 | + 43 | + 18 | + 8 | = | |

| #16 | 59 | + 21 | + 68 | + 73 | + 97 | + 11 | + 47 | = | |

| #17 | 44 | + 62 | + 66 | + 91 | + 68 | + 41 | + 34 | = | |

| #18 | 100 | + 7 | + 20 | + 69 | + 82 | + 8 | + 49 | = | |

| #19 | 40 | + 83 | + 78 | + 19 | + 83 | + 60 | + 54 | = | |

| #20 | 110 | + 65 | + 63 | + 12 | + 31 | + 63 | + 61 | = | |

| #21 | 84 | + 31 | + 66 | + 52 | + 25 | + 5 | + 70 | = | |

| #22 | 149 | + 95 | + 43 | + 77 | + 92 | + 59 | + 38 | = | |

| #23 | 142 | + 92 | + 27 | + 58 | + 79 | + 83 | + 46 | = | |

| #24 | 119 | + 11 | + 11 | + 10 | + 22 | + 48 | + 34 | = | |

| #25 | 148 | + 35 | + 39 | + 62 | + 24 | + 34 | + 6 | = | |

#26	130	+ 58	+ 44	+ 92	+ 28	+ 3	+ 96	=	

#27	115	+ 5	+ 19	+ 92	+ 62	+ 61	+ 68	=	

#28	45	+ 69	+ 42	+ 6	+ 58	+ 88	+ 68	=	

#29	120	+ 25	+ 14	+ 74	+ 83	+ 54	+ 59	=	

#30	100	+ 23	+ 79	+ 8	+ 34	+ 26	+ 52	=	

#31	108	+ 98	+ 7	+ 83	+ 83	+ 17	+ 61	=	

#32	140	+ 73	+ 92	+ 15	+ 33	+ 21	+ 82	=	

#33	122	+ 90	+ 52	+ 68	+ 25	+ 69	+ 52	=	

#34	144	+ 68	+ 60	+ 94	+ 3	+ 46	+ 48	=	

#35	60	+ 24	+ 31	+ 25	+ 67	+ 55	+ 75	=	

#36	110	+ 63	+ 21	+ 67	+ 42	+ 86	+ 31	=	

#37	84	+ 56	+ 8	+ 52	+ 13	+ 77	+ 75	=	

#38	38	+ 26	+ 32	+ 33	+ 92	+ 91	+ 17	=	

| #39 | 76 | + 82 | + 42 | + 26 | + 62 | + 55 | + 69 | = | |

| #40 | 119 | + 63 | + 5 | + 97 | + 97 | + 69 | + 23 | = | |

| #41 | 81 | + 91 | + 66 | + 60 | + 45 | + 10 | + 86 | = | |

| #42 | 128 | + 32 | + 62 | + 36 | + 92 | + 2 | + 8 | = | |

| #43 | 28 | + 97 | + 12 | + 13 | + 51 | + 33 | + 93 | = | |

| #44 | 24 | + 68 | + 24 | + 83 | + 22 | + 61 | + 26 | = | |

| #45 | 97 | + 47 | + 13 | + 21 | + 72 | + 71 | + 36 | = | |

| #46 | 44 | + 88 | + 81 | + 12 | + 16 | + 16 | + 79 | = | |

| #47 | 32 | + 30 | + 95 | + 85 | + 62 | + 31 | + 11 | = | |

| #48 | 120 | + 29 | + 44 | + 89 | + 83 | + 91 | + 46 | = | |

| #49 | 54 | + 47 | + 73 | + 91 | + 62 | + 75 | + 42 | = | |

| #50 | 139 | + 43 | + 29 | + 70 | + 36 | + 56 | + 36 | = | |

Addition and subtraction #51-100

#51	111	- 70	+ 95	- 67	+ 17	- 45	+ 80	=	

#52	130	- 50	- 34	+ 75	- 82	+ 58	+ 19	=	

#53	76	- 52	+ 39	- 27	+ 93	- 98	+ 39	=	

#54	81	- 24	- 32	+ 85	+ 21	+ 72	- 43	=	

#55	92	- 27	+ 20	+ 4	+ 85	- 63	- 91	=	

#56	108	+ 5	+ 89	- 72	- 11	- 81	+ 18	=	

#57	84	+ 42	- 56	- 37	+ 38	+ 61	- 24	=	

#58	92	- 52	+ 35	+ 98	+ 93	- 51	- 82	=	

#59	59	+ 60	+ 79	- 43	- 23	- 72	+ 30	=	

#60	133	+ 47	- 2	- 56	+ 87	- 96	+ 44	=	

#61	150	- 99	+ 17	+ 54	- 80	+ 16	- 34	=	

#62	81	- 52	+ 99	+ 86	+ 26	- 55	- 24	=	

#63	80	- 37	+ 69	+ 49	- 39	+ 26	- 38	=	

#64	81	- 35	+ 72	- 20	+ 66	- 61	+ 91	=	

#65	127	+ 62	- 67	- 58	- 40	+ 6	+ 2	=	

#66	23	+ 23	+ 50	+ 50	- 47	- 37	- 40	=	

#67	108	- 87	+ 43	+ 16	+ 8	- 18	- 33	=	

#68	56	- 2	+ 70	+ 12	- 56	- 17	+ 94	=	

#69	29	+ 23	+ 39	- 54	+ 44	- 27	- 2	=	

#70	47	+ 85	+ 42	- 23	+ 71	- 17	- 18	=	

#71	90	+ 24	- 19	- 24	+ 67	+ 20	- 82	=	

#72	129	+ 18	- 90	- 40	+ 88	+ 21	- 84	=	

#73	119	- 94	+ 58	+ 75	+ 6	- 18	- 27	=	

#74	137	+ 19	+ 54	- 66	- 94	+ 4	- 25	=	

#75	99	- 77	+ 49	+ 97	+ 35	- 32	- 58	=	

#76	122	- 57	- 9	- 24	+ 33	+ 72	+ 25	=	

#77	45	+ 85	- 32	+ 31	+ 17	- 2	- 83	=	

#78	49	+ 72	+ 85	- 98	- 60	+ 70	- 22	=	

#79	81	- 20	+ 33	- 9	- 39	+ 21	+ 87	=	

#80	106	- 77	+ 58	+ 82	+ 17	- 69	- 76	=	

#81	67	+ 65	- 77	- 37	+ 26	+ 26	- 39	=	

#82	68	+ 76	+ 37	+ 98	- 53	- 18	- 85	=	

#83	34	+ 54	- 26	- 35	+ 42	- 4	+ 75	=	

#84	58	- 29	+ 89	- 80	+ 8	+ 8	- 34	=	

#85	44	+ 37	+ 64	+ 38	- 55	- 5	- 95	=	

#86	132	- 56	- 20	- 27	+ 26	+ 10	+ 33	=	

#87	78	- 2	- 57	+ 74	- 69	+ 67	+ 9	=	

#88	92	+ 57	+ 56	- 36	- 14	+ 45	- 92	=	

#89	125	+ 83	- 20	+ 30	+ 28	- 22	- 18	=	

#90	109	+ 59	- 52	+ 32	- 84	+ 52	- 80	=	

#91	66	+ 30	+ 48	- 33	+ 93	- 80	- 32	=	

#92	141	- 88	+ 11	- 35	+ 75	- 93	+ 94	=	

#93	96	- 31	+ 93	- 97	+ 53	- 25	+ 58	=	

#94	131	+ 27	+ 52	- 14	- 6	- 53	+ 53	=	

#95	123	+ 50	+ 92	+ 13	- 92	- 91	- 28	=	

#96	79	+ 94	+ 43	- 28	- 34	+ 22	- 96	=	

#97	90	- 27	+ 98	- 99	- 27	+ 85	+ 7	=	

#98	38	+ 71	+ 35	- 46	- 17	- 38	+ 35	=	

#99	147	- 72	- 20	+ 31	- 37	+ 24	+ 23	=	

#100	45	+ 8	+ 64	- 47	+ 34	- 29	- 3	=	

Addition, subtraction, multiplication #101-150

#101	55	- 34	x 4	x 7	- 20	+ 14	+ 94	=	

#102	99	- 54	x 3	- 84	x 8	+ 49	+ 23	=	

#103	54	- 32	x 3	x 6	+ 90	- 99	+ 68	=	

#104	94	- 68	x 3	- 65	+ 3	x 5	+ 10	=	

#105	25	x 5	- 90	x 8	+ 95	- 28	+ 5	=	

#106	131	+ 27	- 80	- 66	x 4	x 4	+ 47	=	

#107	84	+ 20	- 89	x 7	- 90	x 9	+ 5	=	

#108	46	x 4	- 52	- 52	x 6	+ 38	+ 53	=	

#109	70	- 57	+ 13	x 4	- 8	x 5	+ 71	=	

#110	47	+ 8	- 38	+ 7	x 7	- 88	x 9	=	

#111	50	x 3	- 65	x 6	+ 13	+ 51	- 54	=	

#112	54	- 25	+ 65	- 82	x 5	x 6	+ 56	=	

#113	93	- 69	x 4	x 6	- 29	+ 48	+ 96	=	

#114	29	+ 82	- 96	x 5	+ 23	x 8	- 89	=	

#115	119	- 91	x 3	x 5	- 34	+ 45	+ 54	=	

#116	130	+ 5	- 95	x 4	- 80	x 3	+ 68	=	

#117	98	- 82	x 4	x 8	- 51	+ 17	+ 72	=	

#118	77	x 3	- 95	+ 2	- 68	+ 23	x 4	=	

#119	43	+ 64	+ 13	- 99	x 3	x 3	- 45	=	

#120	52	- 37	x 6	- 40	+ 46	x 9	+ 2	=	

#121	52	- 22	x 4	- 28	x 4	+ 41	+ 48	=	

#122	117	- 79	x 4	- 71	x 3	+ 87	+ 82	=	

#123	38	x 3	- 44	- 29	+ 14	x 7	+ 58	=	

#124	60	x 3	- 68	- 93	+ 3	x 8	+ 39	=	

#125	65	- 27	x 3	- 94	x 6	+ 34	+ 86	=	

#126	148	- 48	+ 6	+ 2	- 90	x 3	x 4	=	

#127	84	- 68	x 5	x 5	+ 17	+ 38	- 63	=	

#128	62	x 3	+ 32	- 72	- 48	x 9	+ 9	=	

#129	109	+ 5	- 99	x 3	x 6	- 32	+ 5	=	

#130	121	- 98	+ 68	- 58	x 3	x 8	+ 25	=	

#131	21	x 6	- 26	- 42	x 3	+ 77	+ 43	=	

#132	115	- 93	x 6	- 47	x 7	+ 36	+ 18	=	

#133	149	+ 5	- 44	- 94	x 3	x 5	+ 59	=	

#134	38	x 3	+ 29	- 47	x 9	- 62	+ 22	=	

#135	113	+ 20	+ 21	- 89	x 3	- 97	x 3	=	

#136	32	+ 40	+ 87	- 45	- 88	x 3	x 9	=	

#137	147	- 93	- 24	x 3	+ 4	x 8	+ 68	=	

#138	98	+ 16	+ 3	- 11	- 94	x 6	x 5	=	

#139	51	x 3	- 58	x 5	+ 51	- 27	+ 28	=	

#140	60	x 3	+ 45	- 93	- 37	x 7	+ 43	=	

#141	68	- 54	x 5	x 5	+ 27	+ 53	- 21	=	

#142	71	+ 33	- 77	x 4	- 23	x 6	+ 10	=	

#143	36	x 3	- 2	+ 28	+ 6	- 44	x 9	=	

#144	121	- 96	x 3	- 56	+ 43	x 3	+ 11	=	

#145	97	+ 11	- 83	x 3	x 4	+ 81	- 33	=	

#146	130	- 92	x 4	- 90	x 8	+ 63	+ 25	=	

#147	111	- 76	x 4	- 93	+ 11	x 4	+ 42	=	

#148	126	- 88	x 4	- 65	+ 7	x 6	+ 10	=	

#149	51	- 27	x 3	- 27	x 4	+ 26	+ 41	=	

#150	126	- 86	x 3	+ 27	- 81	x 4	+ 62	=	

Addition, subtraction, multiplication, division #151-200

| #151 | 20 | x 3 | x 6 | : 30 | + 28 | + 72 | : 56 | = | |

| #152 | 173 | + 54 | + 63 | : 29 | x 6 | x 5 | : 5 | = | |

| #153 | 120 | + 49 | : 13 | + 42 | x 8 | : 20 | x 7 | = | |

| #154 | 130 | : 13 | x 4 | x 4 | : 16 | + 9 | + 4 | = | |

| #155 | 173 | + 22 | : 5 | x 5 | : 13 | + 75 | x 4 | = | |

| #156 | 176 | : 11 | x 4 | x 6 | + 31 | : 5 | + 76 | = | |

| #157 | 80 | x 9 | - 40 | : 8 | x 9 | - 3 | : 2 | = | |

| #158 | 62 | x 6 | : 31 | + 44 | x 4 | : 32 | + 26 | = | |

| #159 | 163 | + 97 | : 5 | x 3 | - 54 | : 17 | x 7 | = | |

| #160 | 169 | + 19 | : 47 | x 3 | x 9 | : 36 | + 10 | = | |

| #161 | 82 | - 71 | + 3 | x 8 | : 4 | x 6 | : 2 | = | |

| #162 | 45 | x 8 | : 10 | : 9 | x 5 | + 86 | - 86 | = | |

| #163 | 22 | + 59 | x 5 | : 5 | : 27 | + 15 | x 5 | = | |

| #164 | 177 | : 59 | + 5 | x 4 | : 16 | + 68 | x 6 | = | |

| #165 | 79 | + 23 | - 34 | x 9 | : 4 | : 17 | x 6 | = | |

| #166 | 90 | x 8 | : 30 | x 8 | - 89 | + 19 | : 2 | = | |

| #167 | 91 | x 3 | : 21 | + 42 | - 39 | x 7 | : 14 | = | |

| #168 | 77 | : 7 | + 61 | x 4 | : 9 | x 9 | - 93 | = | |

| #169 | 193 | + 80 | - 9 | : 33 | x 9 | x 6 | : 3 | = | |

| #170 | 114 | - 59 | x 3 | + 87 | : 14 | x 9 | : 54 | = | |

| #171 | 181 | + 71 | : 42 | x 3 | x 9 | : 54 | + 41 | = | |

| #172 | 119 | - 71 | x 3 | : 8 | x 4 | + 52 | : 2 | = | |

| #173 | 67 | + 13 | : 5 | x 4 | x 9 | + 36 | : 3 | = | |

| #174 | 56 | x 7 | : 56 | + 67 | x 6 | - 46 | : 2 | = | |

| #175 | 50 | x 6 | : 4 | + 41 | : 2 | x 3 | + 13 | = | |

| #176 | 181 | + 69 | : 25 | + 56 | x 9 | : 33 | x 4 | = | |

| #177 | 108 | : 12 | x 8 | x 9 | - 93 | + 55 | : 2 | = | |

| #178 | 156 | : 52 | + 84 | x 6 | : 18 | + 36 | x 4 | = | |

| #179 | 45 | x 7 | - 13 | - 71 | : 3 | x 9 | : 11 | = | |

| #180 | 124 | : 31 | x 3 | + 2 | x 8 | : 4 | + 33 | = | |

| #181 | 179 | + 53 | : 58 | x 5 | x 8 | + 88 | : 8 | = | |

| #182 | 98 | x 7 | : 7 | + 27 | : 5 | + 29 | x 6 | = | |

| #183 | 93 | x 8 | : 6 | : 31 | + 8 | x 7 | + 85 | = | |

| #184 | 182 | + 93 | : 11 | x 9 | : 5 | x 8 | - 75 | = | |

| #185 | 59 | + 4 | : 7 | x 9 | + 59 | : 10 | x 7 | = | |

| #186 | 66 | x 4 | : 3 | - 36 | x 8 | - 8 | : 3 | = | |

| #187 | 195 | : 13 | x 9 | : 15 | x 9 | - 14 | + 60 | = | |

| #188 | 55 | : 5 | + 19 | x 3 | x 6 | : 45 | + 58 | = | |

| #189 | 139 | + 14 | : 17 | x 5 | + 43 | : 2 | x 8 | = | |

| #190 | 154 | : 7 | x 4 | x 4 | + 28 | : 5 | - 43 | = | |

| #191 | 161 | - 96 | - 29 | x 8 | : 4 | : 12 | x 9 | = | |

| #192 | 158 | - 94 | : 8 | x 4 | : 8 | + 91 | x 6 | = | |

| #193 | 98 | : 7 | x 9 | : 7 | x 6 | - 96 | + 71 | = | |

| #194 | 36 | x 5 | : 60 | + 29 | x 8 | : 2 | - 13 | = | |

| #195 | 78 | - 54 | x 4 | : 8 | + 13 | x 5 | : 5 | = | |

| #196 | 158 | : 2 | + 70 | + 59 | : 52 | x 6 | x 9 | = | |

| #197 | 130 | - 82 | : 12 | x 7 | + 42 | x 5 | : 10 | = | |

| #198 | 137 | + 88 | : 25 | x 6 | + 18 | x 6 | : 48 | = | |

| #199 | 90 | x 8 | : 18 | x 5 | : 25 | + 49 | + 43 | = | |

| #200 | 63 | + 9 | + 60 | : 2 | x 4 | : 33 | x 4 | = | |

Addition, subtraction, multiplication, division, fractions #201-250

| #201 | 59 | + 27 | x 8 | - 19 | 4/6 of it | : 2 | + 93 | = | |

| #202 | 130 | - 18 | + 67 | + 28 | : 3 | 4/6 of it | x 9 | = | |

| #203 | 149 | + 13 | : 6 | x 8 | - 65 | + 99 | 4/5 of it | = | |

| #204 | 67 | + 41 | 2/3 of it | - 52 | x 7 | : 28 | + 52 | = | |

| #205 | 121 | + 37 | : 2 | + 7 | - 66 | x 7 | 6/8 of it | = | |

| #206 | 42 | + 99 | 2/3 of it | x 5 | - 63 | + 90 | : 7 | = | |

| #207 | 85 | x 3 | 1/5 of it | : 17 | + 79 | - 69 | + 96 | = | |

| #208 | 135 | - 78 | x 4 | 1/4 of it | : 19 | + 2 | + 62 | = | |

#209	121	: 11	+ 85	- 56	5/8 of it	x 4	+ 37	=	

#210	111	4/6 of it	: 2	+ 14	- 26	x 9	+ 6	=	

#211	56	1/4 of it	+ 8	x 8	+ 58	- 15	: 3	=	

#212	108	- 97	+ 80	x 3	3/7 of it	: 39	+ 24	=	

#213	119	2/7 of it	x 8	- 78	: 2	+ 75	- 79	=	

#214	126	- 91	: 5	+ 65	+ 98	1/5 of it	x 6	=	

#215	141	+ 89	: 10	+ 95	- 94	6/8 of it	x 7	=	

#216	27	+ 75	: 3	x 5	- 70	1/5 of it	+ 95	=	

#217	92	: 4	+ 36	+ 3	x 4	6/8 of it	- 50	=	

#218	137	+ 16	- 58	: 19	+ 30	2/7 of it	x 3	=	

#219	85	- 19	x 5	4/6 of it	- 11	: 19	+ 32	=	

#220	144	: 8	+ 51	- 40	+ 81	4/5 of it	x 4	=	

#221	45	x 9	2/5 of it	- 44	: 2	+ 97	- 55	=	

#222	93	4/6 of it	- 36	x 6	+ 50	: 2	+ 79	=	

#223	54	: 6	square it	2/3 of it	- 27	x 9	- 42	=	

#224	98	x 3	5/6 of it	: 35	+ 16	+ 69	- 2	=	

#225	126	+ 30	: 4	+ 6	3/5 of it	x 4	- 4	=	

#226	75	: 25	+ 60	2/7 of it	x 5	- 78	+ 8	=	

22

| #227 | 129 | + 19 | - 66 | : 41 | + 48 | x 4 | 2/5 of it | = | |

| #228 | 70 | x 7 | 4/5 of it | - 51 | + 62 | : 13 | + 7 | = | |

| #229 | 77 | 5/7 of it | x 6 | - 27 | - 16 | : 41 | + 86 | = | |

| #230 | 44 | + 48 | x 9 | 1/4 of it | - 47 | : 10 | + 12 | = | |

| #231 | 53 | + 99 | - 68 | 4/6 of it | : 4 | x 9 | - 83 | = | |

| #232 | 115 | - 67 | : 8 | + 65 | + 55 | 4/7 of it | x 6 | = | |

| #233 | 30 | x 7 | - 94 | 2/8 of it | + 87 | - 2 | : 57 | = | |

| #234 | 147 | 2/7 of it | x 5 | - 64 | - 49 | + 25 | : 2 | = | |

| #235 | 35 | : 5 | + 81 | x 7 | 7/8 of it | - 91 | + 30 | = | |

#236	25	+ 81	: 53	+ 70	- 27	3/5 of it	x 8	=	

#237	36	x 3	+ 61	- 73	: 6	+ 94	1/5 of it	=	

#238	140	+ 26	- 62	: 8	+ 91	5/8 of it	x 4	=	

#239	70	x 3	+ 74	- 8	: 3	+ 60	5/8 of it	=	

#240	24	1/6 of it	x 3	square it	+ 68	: 4	+ 36	=	

#241	105	- 44	+ 56	: 3	x 3	+ 75	7/8 of it	=	

#242	113	+ 70	- 82	+ 59	2/8 of it	: 10	x 4	=	

#243	116	6/8 of it	- 53	+ 82	: 29	square it	x 3	=	

#244	43	+ 11	x 3	- 74	3/8 of it	: 11	+ 96	=	

								=	
#245	134	- 77	x 8	1/8 of it	- 21	+ 26	: 2	=	

								=	
#246	75	- 34	+ 69	: 55	+ 19	x 3	1/7 of it	=	

								=	
#247	139	+ 24	+ 73	: 2	- 73	x 3	2/3 of it	=	

								=	
#248	47	+ 53	4/5 of it	: 4	+ 6	x 5	- 71	=	

								=	
#249	114	: 6	+ 3	x 8	- 24	5/8 of it	- 84	=	

								=	
#250	97	+ 24	: 11	+ 51	x 6	5/6 of it	- 81	=	

Addition, subtraction, multiplication, division, percentages #251-300

| #251 | 58 | + 18 | x 6 | 50% of it | + 4 | - 50 | : 7 | = | |

| #252 | 95 | x 9 | + 92 | + 14 | - 46 | : 3 | 80% of it | = | |

| #253 | 49 | x 9 | - 45 | - 37 | + 36 | 40% of it | : 2 | = | |

| #254 | 111 | - 33 | - 24 | 50% of it | x 5 | : 5 | + 23 | = | |

| #255 | 131 | + 24 | - 85 | 80% of it | x 3 | : 24 | + 22 | = | |

| #256 | 90 | 70% of it | - 19 | x 7 | + 37 | : 23 | square it | = | |

| #257 | 70 | 30% of it | x 4 | : 14 | + 61 | + 99 | - 63 | = | |

| #258 | 110 | 70% of it | x 7 | + 27 | + 36 | : 2 | - 69 | = | |

#259	69	: 23	+ 96	- 23	+ 24	20% of it	x 4	=

#260	85	: 5	+ 73	x 4	40% of it	- 32	- 75	=

#261	110	60% of it	: 2	x 4	+ 82	- 45	+ 17	=

#262	106	- 54	: 26	+ 56	- 24	x 5	50% of it	=

#263	60	- 20	40% of it	+ 89	: 5	x 4	+ 87	=

#264	117	: 13	x 4	+ 18	50% of it	+ 51	- 4	=

#265	100	+ 82	50% of it	+ 4	- 61	x 4	: 4	=

#266	74	- 38	50% of it	+ 69	x 4	: 12	+ 16	=

#267	98	x 9	- 33	: 3	+ 99	50% of it	+ 3	=

#								
#268	24	50% of it	+ 4	x 7	- 2	+ 26	: 4	=

#								
#269	125	: 5	80% of it	+ 6	x 7	- 93	+ 79	=

#								
#270	73	+ 23	: 4	x 6	50% of it	- 41	+ 27	=

#								
#271	143	- 98	80% of it	x 4	+ 69	: 3	+ 17	=

#								
#272	94	x 7	- 6	+ 18	80% of it	+ 96	: 4	=

#								
#273	45	20% of it	x 5	: 3	+ 97	+ 50	- 76	=

#								
#274	105	40% of it	x 5	+ 83	+ 16	- 4	: 5	=

#								
#275	71	+ 67	- 68	40% of it	x 5	: 7	+ 5	=

#								
#276	94	- 24	: 10	+ 8	x 9	40% of it	+ 92	=

| #277 | 141 | : 47 | + 7 | square it | - 55 | x 9 | 20% of it | = | |

| #278 | 78 | - 13 | 40% of it | x 8 | : 26 | square it | - 50 | = | |

| #279 | 93 | + 29 | + 34 | 50% of it | x 7 | - 7 | : 7 | = | |

| #280 | 82 | - 66 | x 3 | : 8 | + 18 | + 61 | 80% of it | = | |

| #281 | 100 | : 2 | 30% of it | + 26 | + 3 | x 7 | - 74 | = | |

| #282 | 75 | x 7 | : 35 | square it | - 36 | + 7 | 50% of it | = | |

| #283 | 60 | - 25 | 40% of it | x 8 | + 64 | - 7 | : 13 | = | |

| #284 | 36 | : 18 | + 24 | + 69 | - 43 | x 5 | 30% of it | = | |

| #285 | 31 | + 73 | - 49 | 80% of it | x 7 | - 16 | : 2 | = | |

#286	85	x 6	30% of it	- 96	: 19	+ 16	+ 72	=	

#287	125	+ 48	+ 63	50% of it	- 60	x 8	: 16	=	

#288	108	- 49	+ 58	- 85	: 4	x 5	80% of it	=	

#289	54	- 26	50% of it	+ 59	+ 35	: 27	x 5	=	

#290	61	+ 94	60% of it	x 7	- 15	: 6	- 87	=	

#291	62	+ 23	x 6	- 92	: 22	+ 6	80% of it	=	

#292	132	- 81	+ 60	- 96	x 8	: 2	30% of it	=	

#293	145	40% of it	x 3	: 29	+ 51	- 24	+ 2	=	

#294	62	+ 90	: 19	x 8	- 32	+ 92	50% of it	=	

#295	98	x 6	50% of it	- 7	: 7	+ 69	- 17	=	

#296	74	: 37	+ 88	x 4	+ 95	80% of it	- 63	=	

#297	56	- 17	x 7	+ 66	: 3	+ 7	40% of it	=	

#298	53	+ 77	40% of it	x 7	- 60	- 22	: 47	=	

#299	100	: 5	x 8	30% of it	+ 63	+ 77	- 6	=	

#300	58	- 26	: 2	x 7	50% of it	- 39	+ 89	=	

Addition, subtraction, multiplication, division, fractions, percentages, increasing difficulty #301-600

| #301 | 94 | - 78 | x 5 | 10% of it | square it | 1/4 of it | + 80 | = | |

| #302 | 75 | 4/6 of it | x 8 | + 45 | 60% of it | - 64 | : 7 | = | |

| #303 | 100 | : 25 | + 61 | 40% of it | x 7 | - 32 | 4/6 of it | = | |

| #304 | 59 | + 70 | - 79 | 20% of it | x 7 | 4/5 of it | : 7 | = | |

| #305 | 52 | 50% of it | x 7 | - 78 | 3/8 of it | : 3 | + 55 | = | |

| #306 | 74 | x 6 | : 3 | - 73 | 4/6 of it | 10% of it | + 72 | = | |

| #307 | 82 | 50% of it | + 69 | 2/5 of it | : 11 | square it | x 5 | = | |

| #308 | 82 | x 5 | : 5 | 50% of it | + 95 | 7/8 of it | - 35 | = | |

								=	
#309	76	x 7	3/7 of it	- 46	+ 93	: 5	60% of it	=	
#310	100	50% of it	+ 69	- 47	: 9	x 9	1/8 of it	=	
#311	82	x 9	50% of it	- 44	+ 95	: 6	5/7 of it	=	
#312	102	5/6 of it	20% of it	+ 67	: 21	square it	x 3	=	
#313	65	80% of it	- 25	2/3 of it	x 7	: 7	+ 60	=	
#314	102	: 2	x 3	- 104	1/7 of it	+ 78	80% of it	=	
#315	80	: 5	x 9	50% of it	- 60	square it	3/8 of it	=	
#316	70	4/7 of it	: 10	+ 74	- 64	x 5	50% of it	=	
#317	58	- 30	1/4 of it	+ 44	x 5	20% of it	: 3	=	

#318	99	+ 69	6/7 of it	50% of it	: 9	square it	x 9	=

#319	59	+ 66	- 89	50% of it	x 4	: 3	1/6 of it	=

#320	107	+ 81	50% of it	x 4	- 72	6/8 of it	: 3	=

#321	71	+ 90	- 93	50% of it	x 4	1/4 of it	: 17	=

#322	92	- 53	: 13	+ 87	x 4	1/6 of it	70% of it	=

#323	74	x 5	70% of it	: 37	+ 109	- 86	5/6 of it	=

#324	97	+ 62	- 79	3/5 of it	: 2	50% of it	square it	=

#325	108	6/8 of it	x 6	50% of it	- 78	: 55	+ 77	=

#326	94	- 45	2/7 of it	x 5	: 14	+ 90	80% of it	=

#327	112	: 28	+ 70	x 3	- 62	60% of it	3/8 of it	=

#328	81	- 36	: 15	+ 69	3/4 of it	50% of it	x 5	=

#329	82	50% of it	+ 51	6/8 of it	x 3	- 53	: 22	=

#330	101	+ 88	- 112	6/7 of it	x 5	: 3	30% of it	=

#331	113	+ 48	6/7 of it	50% of it	x 4	: 23	square it	=

#332	110	50% of it	: 11	+ 60	x 5	2/5 of it	- 112	=

#333	101	+ 70	- 81	3/5 of it	x 3	50% of it	: 27	=

#334	107	+ 67	50% of it	x 5	- 95	1/4 of it	: 17	=

#335	100	10% of it	x 9	1/6 of it	square it	- 66	: 53	=

#336	108	50% of it	- 35	+ 97	: 29	x 6	2/3 of it	**=**

#337	113	+ 69	- 106	x 3	2/3 of it	: 4	50% of it	**=**

#338	76	: 38	+ 108	80% of it	2/8 of it	x 7	- 63	**=**

#339	91	6/7 of it	x 3	+ 76	40% of it	- 109	square it	**=**

#340	109	+ 101	10% of it	2/3 of it	square it	: 4	x 9	**=**

#341	105	2/7 of it	30% of it	x 9	- 62	+ 50	: 3	**=**

#342	101	+ 99	- 86	2/3 of it	50% of it	x 7	: 2	**=**

#343	79	+ 107	5/6 of it	40% of it	x 7	: 2	- 75	**=**

#344	94	x 9	5/6 of it	- 115	20% of it	: 2	+ 118	**=**

#	Start						=	
#345	89	+ 111	60% of it	: 2	- 46	x 3	4/6 of it	=
#346	76	+ 92	- 113	x 8	40% of it	2/8 of it	: 11	=
#347	101	+ 97	- 78	1/8 of it	x 8	10% of it	square it	=
#348	108	- 66	50% of it	6/7 of it	+ 122	: 2	x 6	=
#349	92	50% of it	x 7	5/7 of it	: 23	+ 116	- 78	=
#350	113	+ 94	: 3	x 8	2/3 of it	- 103	80% of it	=
#351	98	- 59	x 5	2/3 of it	30% of it	: 3	+ 76	=
#352	89	+ 114	- 95	: 9	x 10	7/8 of it	20% of it	=
#353	102	: 34	+ 112	80% of it	x 10	1/4 of it	- 94	=

#354	99	: 11	+ 102	- 75	50% of it	x 6	5/6 of it	=

#355	107	+ 70	- 85	x 8	5/8 of it	: 10	50% of it	=

#356	126	2/7 of it	50% of it	x 3	- 21	: 11	+ 116	=

#357	94	- 48	: 2	+ 69	3/4 of it	x 8	50% of it	=

#358	111	2/3 of it	x 10	70% of it	+ 121	- 60	: 3	=

#359	126	50% of it	- 25	x 3	4/6 of it	: 2	+ 106	=

#360	107	+ 64	- 123	2/3 of it	50% of it	x 8	: 16	=

#361	90	1/5 of it	+ 67	- 65	x 4	40% of it	: 16	=

#362	109	+ 66	5/7 of it	20% of it	x 9	- 83	: 71	=

#363	129	2/3 of it	x 3	- 124	50% of it	+ 76	: 11	=

#364	130	20% of it	+ 61	x 10	1/5 of it	- 74	: 20	=

#365	105	: 7	square it	- 124	+ 119	6/8 of it	20% of it	=

#366	111	- 84	x 8	5/8 of it	80% of it	: 12	+ 106	=

#367	112	50% of it	4/7 of it	x 3	: 8	square it	- 95	=

#368	111	: 3	+ 95	3/4 of it	- 71	x 10	40% of it	=

#369	127	+ 86	- 78	: 15	x 9	4/6 of it	50% of it	=

#370	90	5/6 of it	- 24	x 10	20% of it	: 51	+ 74	=

#371	93	- 59	x 5	: 2	20% of it	+ 109	5/7 of it	=

#372	103	+ 65	- 99	x 6	: 2	4/6 of it	50% of it	=	

#373	133	5/7 of it	x 7	20% of it	+ 93	- 131	: 19	=	

#374	111	: 3	+ 95	6/8 of it	x 4	50% of it	- 123	=	

#375	100	1/4 of it	80% of it	x 4	: 8	square it	- 32	=	

#376	87	: 29	+ 67	40% of it	5/7 of it	x 5	- 65	=	

#377	134	50% of it	+ 115	4/7 of it	: 26	square it	x 8	=	

#378	100	6/8 of it	80% of it	: 6	+ 68	x 7	- 96	=	

#379	95	x 5	: 19	20% of it	+ 129	- 20	4/6 of it	=	

#380	129	- 97	x 10	10% of it	3/8 of it	square it	: 18	=	

#381	99	- 73	x 5	: 26	+ 115	30% of it	2/3 of it	=	

#382	95	80% of it	: 4	+ 123	- 84	x 8	1/4 of it	=	

#383	109	+ 119	- 120	: 9	x 10	80% of it	4/6 of it	=	

#384	95	2/5 of it	+ 133	- 87	50% of it	: 7	square it	=	

#385	116	+ 96	- 87	: 5	40% of it	x 8	3/8 of it	=	

#386	124	: 62	+ 114	- 72	x 4	3/4 of it	50% of it	=	

#387	125	: 25	+ 100	80% of it	- 62	x 8	6/8 of it	=	

#388	134	- 91	+ 107	30% of it	: 3	x 4	2/8 of it	=	

#389	97	+ 140	- 123	: 3	x 5	50% of it	4/5 of it	=	

#390	102	50% of it	: 3	+ 137	1/7 of it	x 3	- 27	=	

#391	137	+ 124	: 29	square it	x 5	2/3 of it	60% of it	=	

#392	125	1/5 of it	x 9	- 133	: 2	50% of it	+ 99	=	

#393	126	- 84	5/7 of it	40% of it	+ 140	: 2	x 4	=	

#394	102	4/6 of it	: 4	+ 128	60% of it	x 9	- 128	=	

#395	136	: 8	+ 88	3/7 of it	40% of it	x 4	- 23	=	

#396	110	60% of it	: 3	x 10	1/4 of it	- 38	+ 135	=	

#397	124	: 62	+ 85	- 62	80% of it	x 5	6/8 of it	=	

#398	109	+ 112	- 146	: 5	square it	4/5 of it	20% of it	=	

#399	132	+ 113	1/7 of it	60% of it	x 9	: 21	square it	=	
#400	137	+ 148	80% of it	6/8 of it	: 3	x 3	- 146	=	
#401	102	: 51	+ 140	- 115	x 10	1/6 of it	80% of it	=	
#402	133	- 51	: 41	+ 118	2/8 of it	30% of it	x 10	=	
#403	110	40% of it	: 2	x 4	- 36	2/8 of it	+ 108	=	
#404	141	2/3 of it	x 11	- 142	: 2	+ 89	60% of it	=	
#405	144	: 18	x 6	5/8 of it	10% of it	+ 130	- 92	=	
#406	141	: 3	+ 117	- 116	5/8 of it	60% of it	x 6	=	
#407	128	- 46	: 2	+ 109	4/6 of it	10% of it	x 3	=	

| #408 | 114 | - 94 | + 92 | 7/8 of it | : 2 | x 8 | 50% of it | = | |

| #409 | 144 | 2/8 of it | x 11 | : 33 | + 153 | - 110 | 20% of it | = | |

| #410 | 126 | 4/6 of it | : 14 | square it | x 10 | 60% of it | + 95 | = | |

| #411 | 116 | - 84 | : 8 | square it | + 136 | 50% of it | 6/8 of it | = | |

| #412 | 126 | + 129 | 40% of it | : 2 | - 36 | x 7 | 3/5 of it | = | |

| #413 | 135 | : 9 | x 4 | 40% of it | 1/6 of it | + 101 | - 27 | = | |

| #414 | 115 | : 5 | + 142 | 60% of it | - 75 | 2/8 of it | square it | = | |

| #415 | 126 | 2/7 of it | 50% of it | x 6 | - 20 | : 44 | + 96 | = | |

| #416 | 127 | + 105 | : 2 | 50% of it | x 9 | 4/6 of it | - 113 | = | |

#	Start						=	
#417	121	: 11	+ 144	40% of it	x 4	6/8 of it	- 147	=
#418	147	: 7	5/7 of it	+ 100	80% of it	x 5	- 147	=
#419	115	: 23	+ 130	80% of it	- 87	6/7 of it	x 7	=
#420	124	3/4 of it	- 44	x 5	40% of it	: 49	+ 133	=
#421	149	+ 126	20% of it	- 37	x 7	6/7 of it	: 12	=
#422	112	- 90	x 7	: 77	+ 103	5/7 of it	20% of it	=
#423	158	: 79	+ 122	- 100	x 6	5/6 of it	20% of it	=
#424	120	70% of it	x 7	- 102	+ 119	4/5 of it	: 11	=
#425	145	40% of it	: 2	+ 93	- 104	x 8	1/4 of it	=

| #426 | 128 | 3/4 of it | - 83 | + 143 | : 26 | x 5 | 50% of it | = | |

| #427 | 124 | - 67 | 4/6 of it | : 19 | + 113 | 40% of it | x 5 | = | |

| #428 | 150 | 10% of it | x 10 | - 99 | : 3 | + 102 | 1/7 of it | = | |

| #429 | 145 | : 29 | + 123 | - 58 | 4/7 of it | 40% of it | x 9 | = | |

| #430 | 130 | 70% of it | x 4 | 5/7 of it | : 65 | square it | + 163 | = | |

| #431 | 160 | 1/5 of it | + 163 | - 115 | : 8 | square it | 40% of it | = | |

| #432 | 159 | : 3 | + 147 | 40% of it | x 10 | 1/5 of it | - 132 | = | |

| #433 | 157 | + 138 | 60% of it | - 113 | 2/8 of it | x 11 | : 22 | = | |

| #434 | 118 | - 67 | : 17 | + 97 | 4/5 of it | 80% of it | x 8 | = | |

#								=	
#435	152	5/8 of it	40% of it	: 2	+ 103	- 100	x 11	=	
#436	117	- 68	x 5	60% of it	2/7 of it	+ 141	: 3	=	
#437	134	- 58	1/4 of it	+ 145	: 41	x 10	10% of it	=	
#438	146	: 2	+ 112	60% of it	- 99	square it	1/4 of it	=	
#439	148	1/4 of it	+ 153	50% of it	- 27	x 4	: 8	=	
#440	143	: 11	+ 141	- 140	x 10	80% of it	5/7 of it	=	
#441	152	7/8 of it	- 83	30% of it	x 10	: 75	+ 154	=	
#442	132	- 108	50% of it	x 4	: 16	+ 126	2/3 of it	=	
#443	138	50% of it	- 50	+ 111	: 5	x 4	3/4 of it	=	

| #444 | 149 | + 130 | : 31 | x 11 | - 36 | 5/7 of it | 40% of it | = | |

| #445 | 130 | 1/5 of it | + 119 | 80% of it | - 68 | : 12 | x 10 | = | |

| #446 | 160 | 1/4 of it | 30% of it | x 10 | : 10 | square it | + 139 | = | |

| #447 | 162 | : 81 | + 140 | - 52 | 30% of it | 2/3 of it | x 7 | = | |

| #448 | 168 | : 28 | x 11 | - 35 | + 145 | 5/8 of it | 70% of it | = | |

| #449 | 163 | + 107 | 80% of it | 3/4 of it | : 6 | x 5 | - 105 | = | |

| #450 | 149 | + 148 | : 3 | - 29 | 5/7 of it | 20% of it | x 8 | = | |

| #451 | 170 | 80% of it | - 30 | : 53 | + 126 | 1/8 of it | x 6 | = | |

| #452 | 143 | + 136 | : 3 | x 10 | 40% of it | 3/4 of it | - 114 | = | |

#453	143	: 11	+ 151	- 134	70% of it	2/7 of it	x 11	=	
#454	170	70% of it	2/7 of it	x 9	- 149	+ 108	: 53	=	
#455	136	6/8 of it	: 3	50% of it	+ 152	- 153	x 3	=	
#456	136	3/4 of it	- 72	30% of it	square it	: 27	+ 161	=	
#457	158	: 2	+ 130	- 149	60% of it	3/4 of it	x 7	=	
#458	162	: 9	x 7	+ 138	6/8 of it	- 133	20% of it	=	
#459	144	+ 174	- 122	: 14	x 10	70% of it	2/7 of it	=	
#460	173	+ 147	- 168	: 19	x 5	70% of it	2/7 of it	=	
#461	154	- 116	x 10	20% of it	: 19	+ 170	5/6 of it	=	

#462	160	70% of it	: 28	square it	+ 154	1/5 of it	x 6	=	

#463	153	- 76	: 7	+ 164	1/7 of it	40% of it	square it	=	

#464	152	2/8 of it	x 6	: 4	- 38	+ 135	50% of it	=	

#465	169	: 13	+ 159	50% of it	- 65	5/7 of it	x 12	=	

#466	167	+ 127	: 14	5/7 of it	x 5	- 25	80% of it	=	

#467	133	- 88	80% of it	: 6	x 12	3/8 of it	+ 157	=	

#468	146	- 116	60% of it	x 9	: 54	+ 180	2/3 of it	=	

#469	156	+ 169	80% of it	4/5 of it	- 172	x 6	: 12	=	

#470	137	+ 169	50% of it	- 123	4/6 of it	x 10	: 25	=	

#	Start							=	
#471	143	: 13	+ 131	- 117	4/5 of it	x 11	30% of it	=	
#472	149	+ 155	50% of it	6/8 of it	- 60	x 11	: 18	=	
#473	170	: 17	x 10	10% of it	square it	1/5 of it	+ 119	=	
#474	173	+ 136	- 144	20% of it	x 8	2/8 of it	: 33	=	
#475	141	2/3 of it	- 40	50% of it	x 6	: 81	+ 146	=	
#476	141	: 3	+ 154	- 151	x 4	20% of it	7/8 of it	=	
#477	168	: 14	square it	- 82	50% of it	+ 174	1/5 of it	=	
#478	149	+ 175	: 4	- 26	40% of it	x 4	5/8 of it	=	
#479	151	+ 171	- 157	40% of it	x 3	: 6	4/6 of it	=	

#480	179	+ 148	- 165	5/6 of it	20% of it	x 11	: 99	=	

#481	156	: 39	square it	x 5	20% of it	+ 125	2/3 of it	=	

#482	160	30% of it	: 12	x 4	+ 167	- 155	2/8 of it	=	

#483	146	- 47	: 3	x 12	+ 184	60% of it	3/4 of it	=	

#484	176	: 22	square it	x 7	5/8 of it	70% of it	+ 125	=	

#485	170	30% of it	x 6	1/6 of it	: 3	+ 125	- 102	=	

#486	159	+ 186	80% of it	: 69	square it	x 4	1/4 of it	=	

#487	164	- 131	: 3	+ 189	30% of it	5/6 of it	x 12	=	

#488	145	80% of it	- 81	1/7 of it	+ 181	: 31	x 11	=	

#									
#489	192	7/8 of it	- 148	x 9	20% of it	: 18	+ 156	=	

#490	164	- 135	+ 139	: 6	5/7 of it	x 10	10% of it	=	

#491	169	- 154	square it	2/5 of it	70% of it	: 7	x 7	=	

#492	184	2/8 of it	x 11	- 161	60% of it	: 69	+ 175	=	

#493	192	- 127	: 5	+ 149	50% of it	x 4	5/6 of it	=	

#494	178	- 143	80% of it	x 4	3/7 of it	: 12	square it	=	

#495	164	50% of it	+ 134	: 6	2/8 of it	x 9	- 28	=	

#496	184	50% of it	3/4 of it	: 23	+ 185	- 173	x 12	=	

#497	156	- 108	x 4	5/6 of it	40% of it	: 32	+ 139	=	

#								
#498	195	: 3	- 51	square it	4/7 of it	+ 183	80% of it	=

#								
#499	170	40% of it	- 54	x 6	3/7 of it	: 3	square it	=

#								
#500	166	+ 161	- 186	4/6 of it	x 6	: 2	50% of it	=

#								
#501	176	50% of it	: 4	x 4	1/8 of it	+ 167	- 24	=

#								
#502	171	- 156	x 12	: 12	+ 195	20% of it	6/7 of it	=

#								
#503	198	2/3 of it	50% of it	- 52	x 6	: 12	+ 195	=

#								
#504	195	2/5 of it	x 12	- 164	: 2	50% of it	+ 137	=

#								
#505	181	+ 181	- 137	60% of it	2/5 of it	x 9	: 9	=

#								
#506	172	: 43	+ 161	40% of it	- 39	x 9	2/3 of it	=

#	Start								=	
#507	169	: 13	+ 192	80% of it	6/8 of it	- 51	x 12		=	
#508	153	4/6 of it	50% of it	x 4	: 68	+ 175	- 151		=	
#509	161	: 23	+ 149	2/8 of it	x 11	- 193	50% of it		=	
#510	184	50% of it	x 12	3/4 of it	- 159	: 3	+ 150		=	
#511	195	: 13	x 9	4/5 of it	- 35	+ 142	80% of it		=	
#512	187	: 17	+ 149	2/8 of it	10% of it	square it	x 5		=	
#513	180	30% of it	5/6 of it	: 3	square it	- 195	x 6		=	
#514	184	1/8 of it	+ 169	- 143	x 4	50% of it	: 49		=	
#515	165	- 101	50% of it	: 8	x 11	1/4 of it	+ 157		=	

#516	200	80% of it	6/8 of it	- 51	x 12	: 46	+ 162	=	

#517	202	+ 204	6/7 of it	- 186	50% of it	: 9	x 13	=	

#518	177	- 154	+ 177	3/4 of it	: 6	x 6	80% of it	=	

#519	165	40% of it	x 7	: 22	4/7 of it	+ 162	- 25	=	

#520	180	60% of it	: 2	+ 184	1/7 of it	x 9	- 190	=	

#521	183	2/3 of it	- 90	50% of it	x 12	: 8	+ 161	=	

#522	174	4/6 of it	- 58	x 7	50% of it	: 29	+ 174	=	

#523	208	7/8 of it	50% of it	+ 199	- 210	x 6	: 80	=	

#524	182	+ 150	50% of it	- 54	: 2	1/7 of it	x 11	=	

								=	
#525	206	: 2	+ 142	80% of it	- 182	x 9	6/7 of it	=	
#526	206	+ 179	3/7 of it	60% of it	- 23	: 2	x 12	=	
#527	212	: 4	+ 206	- 149	20% of it	x 4	2/8 of it	=	
#528	193	+ 212	: 9	2/5 of it	x 13	50% of it	- 55	=	
#529	182	1/7 of it	x 13	50% of it	- 136	: 11	+ 161	=	
#530	192	6/8 of it	: 9	x 8	+ 172	60% of it	- 159	=	
#531	174	- 87	x 11	4/6 of it	50% of it	: 11	+ 177	=	
#532	181	+ 206	: 3	- 78	2/3 of it	x 5	50% of it	=	
#533	184	- 148	x 8	: 8	1/4 of it	+ 196	40% of it	=	

57

#534	203	2/7 of it	x 13	50% of it	- 208	: 13	+ 189	=

#535	197	+ 211	2/3 of it	- 206	x 13	50% of it	: 39	=

#536	215	2/5 of it	+ 169	40% of it	: 2	x 9	- 211	=

#537	201	- 157	: 11	+ 184	50% of it	x 4	7/8 of it	=

#538	183	2/3 of it	50% of it	+ 187	: 62	square it	x 9	=

#539	196	: 14	x 5	+ 215	3/5 of it	- 29	50% of it	=

#540	212	: 2	- 71	x 13	4/7 of it	40% of it	+ 174	=

#541	218	- 193	2/5 of it	square it	: 4	80% of it	x 7	=

#542	175	5/7 of it	: 25	+ 179	50% of it	x 4	- 157	=

| #543 | 214 | -
181 | x 11 | 4/6
of it | : 22 | +
159 | 80%
of it | = | |

| #544 | 178 | : 89 | +
151 | -
123 | x 12 | 2/3
of it | 20%
of it | = | |

| #545 | 196 | 2/8
of it | x 13 | -
219 | 50%
of it | : 11 | +
184 | = | |

| #546 | 186 | 50%
of it | - 61 | : 2 | x 4 | 1/4
of it | +
181 | = | |

| #547 | 181 | +
214 | -
199 | : 14 | square
it | 5/7
of it | 20%
of it | = | |

| #548 | 189 | 2/7
of it | : 2 | +
203 | 20%
of it | x 3 | - 98 | = | |

| #549 | 215 | 40%
of it | x 9 | 1/6
of it | +
167 | : 37 | square
it | = | |

| #550 | 194 | : 97 | +
212 | -
174 | 70%
of it | x 4 | 6/8
of it | = | |

| #551 | 217 | -
197 | x 8 | 70%
of it | 6/7
of it | : 48 | +
168 | = | |

#								
#552	185	60% of it	+ 173	3/4 of it	- 39	: 29	square it	=

#								
#553	200	3/8 of it	80% of it	: 10	x 5	+ 173	- 168	=

#								
#554	198	: 66	+ 163	- 43	2/3 of it	x 5	30% of it	=

#								
#555	208	3/8 of it	- 32	x 10	50% of it	: 10	+ 197	=

#								
#556	201	- 162	4/6 of it	x 3	: 3	50% of it	+ 191	=

#								
#557	193	+ 212	- 162	: 3	x 10	10% of it	4/6 of it	=

#								
#558	210	70% of it	: 7	2/3 of it	x 7	- 40	+ 200	=

#								
#559	185	: 5	+ 209	2/3 of it	50% of it	x 10	- 166	=

#								
#560	218	: 2	+ 216	- 165	30% of it	5/8 of it	x 3	=

#561	217	- 203	x 5	20% of it	square it	2/7 of it	+ 184	=

#562	218	: 2	+ 216	1/5 of it	x 8	50% of it	- 184	=

#563	190	30% of it	2/3 of it	x 5	+ 175	- 175	: 19	=

#564	181	+ 199	: 4	x 8	70% of it	6/8 of it	- 161	=

#565	212	: 53	+ 162	- 115	x 12	50% of it	5/6 of it	=

#566	228	2/3 of it	50% of it	: 19	square it	+ 164	- 90	=

#567	209	: 11	+ 169	50% of it	- 79	x 13	2/3 of it	=

#568	199	+ 185	50% of it	: 32	square it	4/6 of it	x 6	=

#569	232	5/8 of it	60% of it	x 13	: 3	- 180	+ 194	=

| #570 | 230 | 10% of it | + 220 | 4/6 of it | - 141 | x 6 | : 21 | = | |

| #571 | 192 | 50% of it | 3/4 of it | : 18 | square it | + 166 | - 150 | = | |

| #572 | 215 | - 167 | 50% of it | 3/8 of it | x 11 | : 9 | + 228 | = | |

| #573 | 195 | - 181 | x 13 | 4/7 of it | 50% of it | : 26 | + 203 | = | |

| #574 | 233 | + 185 | - 202 | 5/6 of it | 30% of it | : 3 | x 4 | = | |

| #575 | 193 | + 172 | 3/5 of it | - 43 | : 22 | square it | 50% of it | = | |

| #576 | 209 | : 11 | + 233 | - 188 | 50% of it | 1/8 of it | x 12 | = | |

| #577 | 203 | 2/7 of it | : 29 | + 200 | - 42 | 20% of it | x 6 | = | |

| #578 | 202 | : 2 | + 219 | 1/4 of it | 70% of it | x 8 | - 229 | = | |

								=	
#579	192	: 2	1/8 of it	x 14	+ 172	50% of it	- 100	=	
#580	208	2/8 of it	50% of it	x 4	+ 204	- 209	: 3	=	
#581	202	+ 221	- 228	60% of it	: 3	x 10	2/3 of it	=	
#582	198	- 184	square it	1/4 of it	: 7	+ 218	40% of it	=	
#583	222	: 37	x 4	1/6 of it	+ 214	- 193	60% of it	=	
#584	234	4/6 of it	50% of it	: 26	+ 234	- 182	x 9	=	
#585	223	+ 188	2/3 of it	- 202	: 6	x 3	50% of it	=	
#586	222	: 74	+ 204	- 169	x 5	80% of it	1/8 of it	=	
#587	202	50% of it	+ 225	- 214	: 2	x 8	1/8 of it	=	

#								
#588	209	: 11	+ 215	2/3 of it	- 93	x 8	50% of it	=
#589	213	+ 205	: 11	x 3	5/6 of it	60% of it	- 28	=
#590	216	7/8 of it	- 64	20% of it	x 11	: 25	+ 240	=
#591	230	- 184	x 7	3/7 of it	50% of it	: 23	+ 226	=
#592	195	- 175	x 12	50% of it	4/6 of it	: 5	+ 199	=
#593	232	- 194	50% of it	+ 197	3/8 of it	: 9	x 4	=
#594	208	1/8 of it	x 13	: 2	+ 205	50% of it	- 104	=
#595	226	50% of it	+ 177	4/5 of it	: 4	x 5	- 185	=
#596	236	50% of it	- 74	3/4 of it	x 10	: 22	+ 179	=

#597	247	+ 237	50% of it	: 2	- 61	x 14	3/4 of it	=	

#598	222	: 2	- 80	+ 189	2/8 of it	x 12	80% of it	=	

#599	218	50% of it	+ 186	3/5 of it	- 81	: 16	x 7	=	

#600	204	4/6 of it	+ 206	50% of it	: 3	- 42	x 5	=	

Answers

#1. 27 + 26 = 53; 53 + 11 = 64; 64 + 70 = 134; 134 + 91 = 225; 225 + 81 = 306; 306 + 9 = 315;

#2. 39 + 61 = 100; 100 + 40 = 140; 140 + 91 = 231; 231 + 79 = 310; 310 + 23 = 333; 333 + 77 = 410;

#3. 147 + 41 = 188; 188 + 42 = 230; 230 + 45 = 275; 275 + 38 = 313; 313 + 3 = 316; 316 + 48 = 364;

#4. 110 + 84 = 194; 194 + 21 = 215; 215 + 41 = 256; 256 + 99 = 355; 355 + 46 = 401; 401 + 32 = 433;

#5. 62 + 88 = 150; 150 + 27 = 177; 177 + 53 = 230; 230 + 80 = 310; 310 + 82 = 392; 392 + 47 = 439;

#6. 65 + 54 = 119; 119 + 99 = 218; 218 + 31 = 249; 249 + 98 = 347; 347 + 14 = 361; 361 + 67 = 428;

#7. 50 + 41 = 91; 91 + 41 = 132; 132 + 65 = 197; 197 + 90 = 287; 287 + 22 = 309; 309 + 79 = 388;

#8. 108 + 98 = 206; 206 + 60 = 266; 266 + 25 = 291; 291 + 63 = 354; 354 + 36 = 390; 390 + 31 = 421;

#9. 50 + 9 = 59; 59 + 17 = 76; 76 + 80 = 156; 156 + 89 = 245; 245 + 15 = 260; 260 + 54 = 314;

#10. 63 + 68 = 131; 131 + 60 = 191; 191 + 14 = 205; 205 + 30 = 235; 235 + 61 = 296; 296 + 30 = 326;

#11. 55 + 33 = 88; 88 + 35 = 123; 123 + 82 = 205; 205 + 10 = 215; 215 + 74 = 289; 289 + 27 = 316;

#12. 73 + 87 = 160; 160 + 86 = 246; 246 + 70 = 316; 316 + 7 = 323; 323 + 21 = 344; 344 + 96 = 440;

#13. 71 + 33 = 104; 104 + 74 = 178; 178 + 53 = 231; 231 + 90 = 321; 321 + 36 = 357; 357 + 23 = 380;

#14. 138 + 25 = 163; 163 + 98 = 261; 261 + 18 = 279; 279 + 49 = 328; 328 + 29 = 357; 357 + 72 = 429;

#15. 98 + 66 = 164; 164 + 56 = 220; 220 + 3 = 223; 223 + 43 = 266; 266 + 18 = 284; 284 + 8 = 292;

#16. 59 + 21 = 80; 80 + 68 = 148; 148 + 73 = 221; 221 + 97 = 318; 318 + 11 = 329; 329 + 47 = 376;

#17. 44 + 62 = 106; 106 + 66 = 172; 172 + 91 = 263; 263 + 68 = 331; 331 + 41 = 372; 372 + 34 = 406;

#18. 100 + 7 = 107; 107 + 20 = 127; 127 + 69 = 196; 196 + 82 = 278; 278 + 8 = 286; 286 + 49 = 335;

#19. 40 + 83 = 123; 123 + 78 = 201; 201 + 19 = 220; 220 + 83 = 303; 303 + 60 = 363; 363 + 54 = 417;

#20. 110 + 65 = 175; 175 + 63 = 238; 238 + 12 = 250; 250 + 31 = 281; 281 + 63 = 344; 344 + 61 = 405;

#21. 84 + 31 = 115; 115 + 66 = 181; 181 + 52 = 233; 233 + 25 = 258; 258 + 5 = 263; 263 + 70 = 333;

#22. 149 + 95 = 244; 244 + 43 = 287; 287 + 77 = 364; 364 + 92 = 456; 456 + 59 = 515; 515 + 38 = 553;

#23. 142 + 92 = 234; 234 + 27 = 261; 261 + 58 = 319; 319 + 79 = 398; 398 + 83 = 481; 481 + 46 = 527;

#24. 119 + 11 = 130; 130 + 11 = 141; 141 + 10 = 151; 151 + 22 = 173; 173 + 48 = 221; 221 + 34 = 255;

#25. 148 + 35 = 183; 183 + 39 = 222; 222 + 62 = 284; 284 + 24 = 308; 308 + 34 = 342; 342 + 6 = 348;

#26. 130 + 58 = 188, 188 + 44 = 232; 232 + 92 = 324; 324 + 28 = 352; 352 + 3 = 355; 355 + 96 = 451;

#27. 115 + 5 = 120; 120 + 19 = 139; 139 + 92 = 231; 231 + 62 = 293; 293 + 61 = 354; 354 + 68 = 422;

#28. 45 + 69 = 114; 114 + 42 = 156; 156 + 6 = 162; 162 + 58 = 220; 220 + 88 = 308; 308 + 68 = 376;

#29. 120 + 25 = 145; 145 + 14 = 159; 159 + 74 = 233; 233 + 83 = 316; 316 + 54 = 370; 370 + 59 = 429;

#30. 100 + 23 = 123; 123 + 79 = 202; 202 + 8 = 210; 210 + 34 = 244; 244 + 26 = 270; 270 + 52 = 322;

#31. 108 + 98 = 206; 206 + 7 = 213; 213 + 83 = 296; 296 + 83 = 379; 379 + 17 = 396; 396 + 61 = 457;

#32. 140 + 73 = 213; 213 + 92 = 305; 305 + 15 = 320; 320 + 33 = 353; 353 + 21 = 374; 374 + 82 = 456;

#33. 122 + 90 = 212; 212 + 52 = 264; 264 + 68 = 332; 332 + 25 = 357; 357 + 69 = 426; 426 + 52 = 478;

#34. 144 + 68 = 212; 212 + 60 = 272; 272 + 94 = 366; 366 + 3 = 369; 369 + 46 = 415; 415 + 48 = 463;

#35. 60 + 24 = 84; 84 + 31 = 115; 115 + 25 = 140; 140 + 67 = 207; 207 + 55 = 262; 262 + 75 = 337;

#36. 110 + 63 = 173; 173 + 21 = 194; 194 + 67 = 261; 261 + 42 = 303; 303 + 86 = 389; 389 + 31 = 420;

#37. 84 + 56 = 140; 140 + 8 = 148; 148 + 52 = 200; 200 + 13 = 213; 213 + 77 = 290; 290 + 75 = 365;

#38. 38 + 26 = 64; 64 + 32 = 96; 96 + 33 = 129; 129 + 92 = 221; 221 + 91 = 312; 312 + 17 = 329;

#39. 76 + 82 = 158; 158 + 42 = 200; 200 + 26 = 226; 226 + 62 = 288; 288 + 55 = 343; 343 + 69 = 412;

#40. 119 + 63 = 182; 182 + 5 = 187; 187 + 97 = 284; 284 + 97 = 381; 381 + 69 = 450; 450 + 23 = 473;

#41. 81 + 91 = 172; 172 + 66 = 238; 238 + 60 = 298; 298 + 45 = 343; 343 + 10 = 353; 353 + 86 = 439;

#42. 128 + 32 = 160; 160 + 62 = 222; 222 + 36 = 258; 258 + 92 = 350; 350 + 2 = 352; 352 + 8 = 360;

#43. 28 + 97 = 125; 125 + 12 = 137; 137 + 13 = 150; 150 + 51 = 201; 201 + 33 = 234; 234 + 93 = 327;

#44. 24 + 68 = 92; 92 + 24 = 116; 116 + 83 = 199; 199 + 22 = 221; 221 + 61 = 282; 282 + 26 = 308;

#45. 97 + 47 = 144; 144 + 13 = 157; 157 + 21 = 178; 178 + 72 = 250; 250 + 71 = 321; 321 + 36 = 357;

#46. 44 + 88 = 132; 132 + 81 = 213; 213 + 12 = 225; 225 + 16 = 241; 241 + 16 = 257; 257 + 79 = 336;

#47. 32 + 30 = 62; 62 + 95 = 157; 157 + 85 = 242; 242 + 62 = 304; 304 + 31 = 335; 335 + 11 = 346;

#48. 120 + 29 = 149; 149 + 44 = 193; 193 + 89 = 282; 282 + 83 = 365; 365 + 91 = 456; 456 + 46 = 502;

#49. 54 + 47 = 101; 101 + 73 = 174; 174 + 91 = 265; 265 + 62 = 327; 327 + 75 = 402; 402 + 42 = 444;

#50. 139 + 43 = 182; 182 + 29 = 211; 211 + 70 = 281; 281 + 36 = 317; 317 + 56 = 373; 373 + 36 = 409;

#51. 111 - 70 = 41; 41 + 95 = 136; 136 - 67 = 69; 69 + 17 = 86; 86 - 45 = 41; 41 + 80 = 121;

#52. 130 - 50 = 80; 80 - 34 = 46; 46 + 75 = 121; 121 - 82 = 39; 39 + 58 = 97; 97 + 19 = 116;

#53. 76 - 52 = 24; 24 + 39 = 63; 63 - 27 = 36; 36 + 93 = 129; 129 - 98 = 31; 31 + 39 = 70;

#54. 81 - 24 = 57; 57 - 32 = 25; 25 + 85 = 110; 110 + 21 = 131; 131 + 72 = 203; 203 - 43 = 160;

#55. 92 - 27 = 65; 65 + 20 = 85; 85 + 4 = 89; 89 + 85 = 174; 174 - 63 = 111; 111 - 91 = 20;

#56. 108 + 5 = 113; 113 + 89 = 202; 202 - 72 = 130; 130 - 11 = 119; 119 - 81 = 38; 38 + 18 = 56;

#57. 84 + 42 = 126; 126 - 56 = 70; 70 - 37 = 33; 33 + 38 = 71; 71 + 61 = 132; 132 - 24 = 108;

#58. 92 - 52 = 40; 40 + 35 = 75; 75 + 98 = 173; 173 + 93 = 266; 266 - 51 = 215; 215 - 82 = 133;

#59. 59 + 60 = 119; 119 + 79 = 198; 198 - 43 = 155; 155 - 23 = 132; 132 - 72 = 60; 60 + 30 = 90;

#60. 133 + 47 = 180; 180 - 2 = 178; 178 - 56 = 122; 122 + 87 = 209; 209 - 96 = 113; 113 + 44 = 157;

#61. 150 - 99 = 51; 51 + 17 = 68; 68 + 54 = 122; 122 - 80 = 42; 42 + 16 = 58; 58 - 34 = 24;

#62. 81 - 52 = 29; 29 + 99 = 128; 128 + 86 = 214; 214 + 26 = 240; 240 - 55 = 185; 185 - 24 = 161;

#63. 80 - 37 = 43; 43 + 69 = 112; 112 + 49 = 161; 161 - 39 = 122; 122 + 26 = 148; 148 - 38 = 110;

#64. 81 - 35 = 46; 46 + 72 = 118; 118 - 20 = 98; 98 + 66 = 164; 164 - 61 = 103; 103 + 91 = 194;

#65. 127 + 62 = 189; 189 - 67 = 122; 122 - 58 = 64; 64 - 40 = 24; 24 + 6 = 30; 30 + 2 = 32;

#66. 23 + 23 = 46; 46 + 50 = 96; 96 + 50 = 146; 146 - 47 = 99; 99 - 37 = 62; 62 - 40 = 22;

#67. 108 - 87 = 21; 21 + 43 = 64; 64 + 16 = 80; 80 + 8 = 88; 88 - 18 = 70; 70 - 33 = 37;

#68. 56 - 2 = 54; 54 + 70 = 124; 124 + 12 = 136; 136 - 56 = 80; 80 - 17 = 63; 63 + 94 = 157;

#69. 29 + 23 = 52; 52 + 39 = 91; 91 - 54 = 37; 37 + 44 = 81; 81 - 27 = 54; 54 - 2 = 52;

#70. 47 + 85 = 132; 132 + 42 = 174; 174 - 23 = 151; 151 + 71 = 222; 222 - 17 = 205; 205 - 18 = 187;

#71. 90 + 24 = 114; 114 - 19 = 95; 95 - 24 = 71; 71 + 67 = 138; 138 + 20 = 158; 158 - 82 = 76;

#72. 129 + 18 = 147; 147 - 90 = 57; 57 - 40 = 17; 17 + 88 = 105; 105 + 21 = 126; 126 - 84 = 42;

#73. 119 - 94 = 25; 25 + 58 = 83; 83 + 75 = 158; 158 + 6 = 164; 164 - 18 = 146; 146 - 27 = 119;

#74. 137 + 19 = 156; 156 + 54 = 210; 210 - 66 = 144; 144 - 94 = 50; 50 + 4 = 54; 54 - 25 = 29;

#75. 99 - 77 = 22; 22 + 49 = 71; 71 + 97 = 168; 168 + 35 = 203; 203 - 32 = 171; 171 - 58 = 113;

#76. 122 - 57 = 65; 65 - 9 = 56; 56 - 24 = 32; 32 + 33 = 65; 65 + 72 = 137; 137 + 25 = 162;

#77. 45 + 85 = 130; 130 - 32 = 98; 98 + 31 = 129; 129 + 17 = 146; 146 - 2 = 144; 144 - 83 = 61;

#78. 49 + 72 = 121; 121 + 85 = 206; 206 - 98 = 108; 108 - 60 = 48; 48 + 70 = 118; 118 - 22 = 96;

#79. 81 - 20 = 61; 61 + 33 = 94; 94 - 9 = 85; 85 - 39 = 46; 46 + 21 = 67; 67 + 87 = 154;

#80. 106 - 77 = 29; 29 + 58 = 87; 87 + 82 = 169; 169 + 17 = 186; 186 - 69 = 117; 117 - 76 = 41;

#81. 67 + 65 = 132; 132 - 77 = 55; 55 - 37 = 18; 18 + 26 = 44; 44 + 26 = 70; 70 - 39 = 31;

#82. 68 + 76 = 144; 144 + 37 = 181; 181 + 98 = 279; 279 - 53 = 226; 226 - 18 = 208; 208 - 85 = 123;

#83. 34 + 54 = 88; 88 - 26 = 62; 62 - 35 = 27; 27 + 42 = 69; 69 - 4 = 65; 65 + 75 = 140;

#84. 58 - 29 = 29; 29 + 89 = 118; 118 - 80 = 38; 38 + 8 = 46; 46 + 8 = 54; 54 - 34 = 20;

#85. 44 + 37 = 81; 81 + 64 = 145; 145 + 38 = 183; 183 - 55 = 128; 128 - 5 = 123; 123 - 95 = 28;

#86. 132 - 56 = 76; 76 - 20 = 56; 56 - 27 = 29; 29 + 26 = 55; 55 + 10 = 65; 65 + 33 = 98;

#87. 78 - 2 = 76; 76 - 57 = 19; 19 + 74 = 93; 93 - 69 = 24; 24 + 67 = 91; 91 + 9 = 100;

#88. 92 + 57 = 149; 149 + 56 = 205; 205 - 36 = 169; 169 - 14 = 155; 155 + 45 = 200; 200 - 92 = 108;

#89. 125 + 83 = 208; 208 - 20 = 188; 188 + 30 = 218; 218 + 28 = 246; 246 - 22 = 224; 224 - 18 = 206;

#90. 109 + 59 = 168; 168 - 52 = 116; 116 + 32 = 148; 148 - 84 = 64; 64 + 52 = 116; 116 - 80 = 36;

#91. 66 + 30 = 96; 96 + 48 = 144; 144 - 33 = 111; 111 + 93 = 204; 204 - 80 = 124; 124 - 32 = 92;

#92. 141 - 88 = 53; 53 + 11 = 64; 64 - 35 = 29; 29 + 75 = 104; 104 - 93 = 11; 11 + 94 = 105;

#93. 96 - 31 = 65; 65 + 93 = 158; 158 - 97 = 61; 61 + 53 = 114; 114 - 25 = 89; 89 + 58 = 147;

#94. 131 + 27 = 158; 158 + 52 = 210; 210 - 14 = 196; 196 - 6 = 190; 190 - 53 = 137; 137 + 53 = 190;

#95. 123 + 50 = 173; 173 + 92 = 265; 265 + 13 = 278; 278 - 92 = 186; 186 - 91 = 95; 95 - 28 = 67;

#96. 79 + 94 = 173; 173 + 43 = 216; 216 - 28 = 188; 188 - 34 = 154; 154 + 22 = 176; 176 - 96 = 80;

#97. 90 - 27 = 63; 63 + 98 = 161; 161 - 99 = 62; 62 - 27 = 35; 35 + 85 = 120; 120 + 7 = 127;

#98. 38 + 71 = 109; 109 + 35 = 144; 144 - 46 = 98; 98 - 17 = 81; 81 - 38 = 43; 43 + 35 = 78;

#99. 147 - 72 = 75; 75 - 20 = 55; 55 + 31 = 86; 86 - 37 = 49; 49 + 24 = 73; 73 + 23 = 96;

#100. 45 + 8 = 53; 53 + 64 = 117; 117 - 47 = 70; 70 + 34 = 104; 104 - 29 = 75; 75 - 3 = 72;

#101. 55 - 34 = 21; 21 x 4 = 84; 84 x 7 = 588; 588 - 20 = 568; 568 + 14 = 582; 582 + 94 = 676;

#102. 99 - 54 = 45; 45 x 3 = 135; 135 - 84 = 51; 51 x 8 = 408; 408 + 49 = 457; 457 + 23 = 480;

#103. 54 - 32 = 22; 22 x 3 = 66; 66 x 6 = 396; 396 + 90 = 486; 486 - 99 = 387; 387 + 68 = 455;

#104. 94 - 68 = 26; 26 x 3 = 78; 78 - 65 = 13; 13 + 3 = 16; 16 x 5 = 80; 80 + 10 = 90;

#105. 25 x 5 = 125; 125 - 90 = 35; 35 x 8 = 280; 280 + 95 = 375; 375 - 28 = 347; 347 + 5 = 352;

#106. 131 + 27 = 158; 158 - 80 = 78; 78 - 66 = 12; 12 x 4 = 48; 48 x 4 = 192; 192 + 47 = 239;

#107. 84 + 20 = 104; 104 - 89 = 15; 15 x 7 = 105; 105 - 90 = 15; 15 x 9 = 135; 135 + 5 = 140;

#108. 46 x 4 = 184; 184 - 52 = 132; 132 - 52 = 80; 80 x 6 = 480; 480 + 38 = 518; 518 + 53 = 571;

#109. 70 - 57 = 13; 13 + 13 = 26; 26 x 4 = 104; 104 - 8 = 96; 96 x 5 = 480; 480 + 71 = 551;

#110. 47 + 8 = 55; 55 - 38 = 17; 17 + 7 = 24; 24 x 7 = 168; 168 - 88 = 80; 80 x 9 = 720;

#111. 50 x 3 = 150; 150 - 65 = 85; 85 x 6 = 510; 510 + 13 = 523; 523 + 51 = 574; 574 - 54 = 520;

#112. 54 - 25 = 29; 29 + 65 = 94; 94 - 82 = 12; 12 x 5 = 60; 60 x 6 = 360; 360 + 56 = 416;

#113. 93 - 69 = 24; 24 x 4 = 96; 96 x 6 = 576; 576 - 29 = 547; 547 + 48 = 595; 595 + 96 = 691;

#114. 29 + 82 = 111; 111 - 96 = 15; 15 x 5 = 75; 75 + 23 = 98; 98 x 8 = 784; 784 - 89 = 695;

#115. 119 - 91 = 28; 28 x 3 = 84; 84 x 5 = 420; 420 - 34 = 386; 386 + 45 = 431; 431 + 54 = 485;

#116. 130 + 5 = 135; 135 - 95 = 40; 40 x 4 = 160; 160 - 80 = 80; 80 x 3 = 240; 240 + 68 = 308;

#117. 98 - 82 = 16; 16 x 4 = 64; 64 x 8 = 512; 512 - 51 = 461; 461 + 17 = 478; 478 + 72 = 550;

#118. 77 x 3 = 231; 231 - 95 = 136; 136 + 2 = 138; 138 - 68 = 70; 70 + 23 = 93; 93 x 4 = 372;

#119. 43 + 64 = 107; 107 + 13 = 120; 120 - 99 = 21; 21 x 3 = 63; 63 x 3 = 189; 189 - 45 = 144;

#120. 52 - 37 = 15; 15 x 6 = 90; 90 - 40 = 50; 50 + 46 = 96; 96 x 9 = 864; 864 + 2 = 866;

#121. 52 - 22 = 30; 30 x 4 = 120; 120 - 28 = 92; 92 x 4 = 368; 368 + 41 = 409; 409 + 48 = 457;

#122. 117 - 79 = 38; 38 x 4 = 152; 152 - 71 = 81; 81 x 3 = 243; 243 + 87 = 330; 330 + 82 = 412;

#123. 38 x 3 = 114; 114 - 44 = 70; 70 - 29 = 41; 41 + 14 = 55; 55 x 7 = 385; 385 + 58 = 443;

#124. 60 x 3 = 180; 180 - 68 = 112; 112 - 93 = 19; 19 + 3 = 22; 22 x 8 = 176; 176 + 39 = 215;

#125. 65 - 27 = 38; 38 x 3 = 114; 114 - 94 = 20; 20 x 6 = 120; 120 + 34 = 154; 154 + 86 = 240;

#126. 148 - 48 = 100; 100 + 6 = 106; 106 + 2 = 108; 108 - 90 = 18; 18 x 3 = 54; 54 x 4 = 216;

#127. 84 - 68 = 16; 16 x 5 = 80; 80 x 5 = 400; 400 + 17 = 417; 417 + 38 = 455; 455 - 63 = 392;

#128. 62 x 3 = 186; 186 + 32 = 218; 218 - 72 = 146; 146 - 48 = 98; 98 x 9 = 882; 882 + 9 = 891;

#129. 109 + 5 = 114; 114 - 99 = 15; 15 x 3 = 45; 45 x 6 = 270; 270 - 32 = 238; 238 + 5 = 243;

#130. 121 - 98 = 23; 23 + 68 = 91; 91 - 58 = 33; 33 x 3 = 99; 99 x 8 = 792; 792 + 25 = 817;

#131. 21 x 6 = 126; 126 - 26 = 100; 100 - 42 = 58; 58 x 3 = 174; 174 + 77 = 251; 251 + 43 = 294;

#132. 115 - 93 = 22; 22 x 6 = 132; 132 - 47 = 85; 85 x 7 = 595; 595 + 36 = 631; 631 + 18 = 649;

#133. 149 + 5 = 154; 154 - 44 = 110; 110 - 94 = 16; 16 x 3 = 48; 48 x 5 = 240; 240 + 59 = 299;

#134. 38 x 3 = 114; 114 + 29 = 143; 143 - 47 = 96; 96 x 9 = 864; 864 - 62 = 802; 802 + 22 = 824;

#135. 113 + 20 = 133; 133 + 21 = 154; 154 - 89 = 65; 65 x 3 = 195; 195 - 97 = 98; 98 x 3 = 294;

#136. 32 + 40 = 72; 72 + 87 = 159; 159 - 45 = 114; 114 - 88 = 26; 26 x 3 = 78; 78 x 9 = 702;

#137. 147 - 93 = 54; 54 - 24 = 30; 30 x 3 = 90; 90 + 4 = 94; 94 x 8 = 752; 752 + 68 = 820;

#138. 98 + 16 = 114; 114 + 3 = 117; 117 - 11 = 106; 106 - 94 = 12; 12 x 6 = 72; 72 x 5 = 360;

#139. 51 x 3 = 153; 153 - 58 = 95; 95 x 5 = 475; 475 + 51 = 526; 526 - 27 = 499; 499 + 28 = 527;

#140. 60 x 3 = 180; 180 + 45 = 225; 225 - 93 = 132; 132 - 37 = 95; 95 x 7 = 665; 665 + 43 = 708;

#141. 68 - 54 = 14; 14 x 5 = 70; 70 x 5 = 350; 350 + 27 = 377; 377 + 53 = 430; 430 - 21 = 409;

#142. 71 + 33 = 104; 104 - 77 = 27; 27 x 4 = 108; 108 - 23 = 85; 85 x 6 = 510; 510 + 10 = 520;

#143. 36 x 3 = 108; 108 - 2 = 106; 106 + 28 = 134; 134 + 6 = 140; 140 - 44 = 96; 96 x 9 = 864;

#144. 121 - 96 = 25; 25 x 3 = 75; 75 - 56 = 19; 19 + 43 = 62; 62 x 3 = 186; 186 + 11 = 197;

#145. 97 + 11 = 108; 108 - 83 = 25; 25 x 3 = 75; 75 x 4 = 300; 300 + 81 = 381; 381 - 33 = 348;

#146. 130 - 92 = 38; 38 x 4 = 152; 152 - 90 = 62; 62 x 8 = 496; 496 + 63 = 559; 559 + 25 = 584;

#147. 111 - 76 = 35; 35 x 4 = 140; 140 - 93 = 47; 47 + 11 = 58; 58 x 4 = 232; 232 + 42 = 274;

#148. 126 - 88 = 38; 38 x 4 = 152; 152 - 65 = 87; 87 + 7 = 94; 94 x 6 = 564; 564 + 10 = 574;

#149. 51 - 27 = 24; 24 x 3 = 72; 72 - 27 = 45; 45 x 4 = 180; 180 + 26 = 206; 206 + 41 = 247;

#150. 126 - 86 = 40; 40 x 3 = 120; 120 + 27 = 147; 147 - 81 = 66; 66 x 4 = 264; 264 + 62 = 326;

#151. 20 x 3 = 60; 60 x 6 = 360; 360 : 30 = 12; 12 + 28 = 40; 40 + 72 = 112; 112 : 56 = 2;

#152. 173 + 54 = 227; 227 + 63 = 290; 290 : 29 = 10; 10 x 6 = 60; 60 x 5 = 300; 300 : 5 = 60;

#153. 120 + 49 = 169; 169 : 13 = 13; 13 + 42 = 55; 55 x 8 = 440; 440 : 20 = 22; 22 x 7 = 154;

#154. 130 : 13 = 10; 10 x 4 = 40; 40 x 4 = 160; 160 : 16 = 10; 10 + 9 = 19; 19 + 4 = 23;

#155. 173 + 22 = 195; 195 : 5 = 39; 39 x 5 = 195; 195 : 13 = 15; 15 + 75 = 90; 90 x 4 = 360;

#156. 176 : 11 = 16; 16 x 4 = 64; 64 x 6 = 384; 384 + 31 = 415; 415 : 5 = 83; 83 + 76 = 159;

#157. 80 x 9 = 720; 720 - 40 = 680; 680 : 8 = 85; 85 x 9 = 765; 765 - 3 = 762; 762 : 2 = 381;

#158. 62 x 6 = 372; 372 : 31 = 12; 12 + 44 = 56; 56 x 4 = 224; 224 : 32 = 7; 7 + 26 = 33;

#159. 163 + 97 = 260; 260 : 5 = 52; 52 x 3 = 156; 156 - 54 = 102; 102 : 17 = 6; 6 x 7 = 42;

#160. 169 + 19 = 188; 188 : 47 = 4; 4 x 3 = 12; 12 x 9 = 108; 108 : 36 = 3; 3 + 10 = 13;

#161. 82 - 71 = 11; 11 + 3 = 14; 14 x 8 = 112; 112 : 4 = 28; 28 x 6 = 168; 168 : 2 = 84;

#162. 45 x 8 = 360; 360 : 10 = 36; 36 : 9 = 4; 4 x 5 = 20; 20 + 86 = 106; 106 - 86 = 20;

#163. 22 + 59 = 81; 81 x 5 = 405; 405 : 5 = 81; 81 : 27 = 3; 3 + 15 = 18; 18 x 5 = 90;

#164. 177 : 59 = 3; 3 + 5 = 8; 8 x 4 = 32; 32 : 16 = 2; 2 + 68 = 70; 70 x 6 = 420;

#165. 79 + 23 = 102; 102 - 34 = 68; 68 x 9 = 612; 612 : 4 = 153; 153 : 17 = 9; 9 x 6 = 54;

#166. 90 x 8 = 720; 720 : 30 = 24; 24 x 8 = 192; 192 - 89 = 103; 103 + 19 = 122; 122 : 2 = 61;

#167. 91 x 3 = 273; 273 : 21 = 13; 13 + 42 = 55; 55 - 39 = 16; 16 x 7 = 112; 112 : 14 = 8;

#168. 77 : 7 = 11; 11 + 61 = 72; 72 x 4 = 288; 288 : 9 = 32; 32 x 9 = 288; 288 - 93 = 195;

#169. 193 + 80 = 273; 273 - 9 = 264; 264 : 33 = 8; 8 x 9 = 72; 72 x 6 = 432; 432 : 3 = 144;

#170. 114 - 59 = 55; 55 x 3 = 165; 165 + 87 = 252; 252 : 14 = 18; 18 x 9 = 162; 162 : 54 = 3;

#171. 181 + 71 = 252; 252 : 42 = 6; 6 x 3 = 18; 18 x 9 = 162; 162 : 54 = 3; 3 + 41 = 44;

#172. 119 - 71 = 48; 48 x 3 = 144; 144 : 8 = 18; 18 x 4 = 72; 72 + 52 = 124; 124 : 2 = 62;

#173. 67 + 13 = 80; 80 : 5 = 16; 16 x 4 = 64; 64 x 9 = 576; 576 + 36 = 612; 612 : 3 = 204;

#174. 56 x 7 = 392; 392 : 56 = 7; 7 + 67 = 74; 74 x 6 = 444; 444 - 46 = 398; 398 : 2 = 199;

#175. 50 x 6 = 300; 300 : 4 = 75; 75 + 41 = 116; 116 : 2 = 58; 58 x 3 = 174; 174 + 13 = 187;

#176. 181 + 69 = 250; 250 : 25 = 10; 10 + 56 = 66; 66 x 9 = 594; 594 : 33 = 18; 18 x 4 = 72;

#177. 108 : 12 = 9; 9 x 8 = 72; 72 x 9 = 648; 648 - 93 = 555; 555 + 55 = 610; 610 : 2 = 305;

#178. 156 : 52 = 3; 3 + 84 = 87; 87 x 6 = 522; 522 : 18 = 29; 29 + 36 = 65; 65 x 4 = 260;

#179. 45 x 7 = 315; 315 - 13 = 302; 302 - 71 = 231; 231 : 3 = 77; 77 x 9 = 693; 693 : 11 = 63;

#180. 124 : 31 = 4; 4 x 3 = 12; 12 + 2 = 14; 14 x 8 = 112; 112 : 4 = 28; 28 + 33 = 61;

#181. 179 + 53 = 232; 232 : 58 = 4; 4 x 5 = 20; 20 x 8 = 160; 160 + 88 = 248; 248 : 8 = 31;

#182. 98 x 7 = 686; 686 : 7 = 98; 98 + 27 = 125; 125 : 5 = 25; 25 + 29 = 54; 54 x 6 = 324;

#183. 93 x 8 = 744; 744 : 6 = 124; 124 : 31 = 4; 4 + 8 = 12; 12 x 7 = 84; 84 + 85 = 169;

#184. 182 + 93 = 275; 275 : 11 = 25; 25 x 9 = 225; 225 : 5 = 45; 45 x 8 = 360; 360 - 75 = 285;

#185. 59 + 4 = 63; 63 : 7 = 9; 9 x 9 = 81; 81 + 59 = 140; 140 : 10 = 14; 14 x 7 = 98;

#186. 66 x 4 = 264; 264 : 3 = 88; 88 - 36 = 52; 52 x 8 = 416; 416 - 8 = 408; 408 : 3 = 136;

#187. 195 : 13 = 15; 15 x 9 = 135; 135 : 15 = 9; 9 x 9 = 81; 81 - 14 = 67; 67 + 60 = 127;

#188. 55 : 5 = 11; 11 + 19 = 30; 30 x 3 = 90; 90 x 6 = 540; 540 : 45 = 12; 12 + 58 = 70;

#189. 139 + 14 = 153; 153 : 17 = 9; 9 x 5 = 45; 45 + 43 = 88; 88 : 2 = 44; 44 x 8 = 352;

#190. 154 : 7 = 22; 22 x 4 = 88; 88 x 4 = 352; 352 + 28 = 380; 380 : 5 = 76; 76 - 43 = 33;

#191. 161 - 96 = 65; 65 - 29 = 36; 36 x 8 = 288; 288 : 4 = 72; 72 : 12 = 6; 6 x 9 = 54;

#192. 158 - 94 = 64; 64 : 8 = 8; 8 x 4 = 32; 32 : 8 = 4; 4 + 91 = 95; 95 x 6 = 570;

#193. 98 : 7 = 14; 14 x 9 = 126; 126 : 7 = 18; 18 x 6 = 108; 108 - 96 = 12; 12 + 71 = 83;

#194. 36 x 5 = 180; 180 : 60 = 3; 3 + 29 = 32; 32 x 8 = 256; 256 : 2 = 128; 128 - 13 = 115;

#195. 78 - 54 = 24; 24 x 4 = 96; 96 : 8 = 12; 12 + 13 = 25; 25 x 5 = 125; 125 : 5 = 25;

#196. 158 : 2 = 79; 79 + 70 = 149; 149 + 59 = 208; 208 : 52 = 4; 4 x 6 = 24; 24 x 9 = 216;

#197. 130 - 82 = 48; 48 : 12 = 4; 4 x 7 = 28; 28 + 42 = 70; 70 x 5 = 350; 350 : 10 = 35;

#198. 137 + 88 = 225; 225 : 25 = 9; 9 x 6 = 54; 54 + 18 = 72; 72 x 6 = 432; 432 : 48 = 9;

#199. 90 x 8 = 720; 720 : 18 = 40; 40 x 5 = 200; 200 : 25 = 8; 8 + 49 = 57; 57 + 43 = 100;

#200. 63 + 9 = 72; 72 + 60 = 132; 132 : 2 = 66; 66 x 4 = 264; 264 : 33 = 8; 8 x 4 = 32;

#201. 59 + 27 = 86; 86 x 8 = 688; 688 - 19 = 669; 669 (4/6 of it) = 446; 446 : 2 = 223; 223 + 93 = 316;

#202. 130 - 18 = 112; 112 + 67 = 179; 179 + 28 = 207; 207 : 3 = 69; 69 (4/6 of it) = 46; 46 x 9 = 414;

#203. 149 + 13 = 162; 162 : 6 = 27; 27 x 8 = 216; 216 - 65 = 151; 151 + 99 = 250; 250 (4/5 of it) = 200;

#204. 67 + 41 = 108; 108 (2/3 of it) = 72; 72 - 52 = 20; 20 x 7 = 140; 140 : 28 = 5; 5 + 52 = 57;

#205. 121 + 37 = 158; 158 : 2 = 79; 79 + 7 = 86; 86 - 66 = 20; 20 x 7 = 140; 140 (6/8 of it) = 105;

#206. 42 + 99 = 141; 141 (2/3 of it) = 94; 94 x 5 = 470; 470 - 63 = 407; 407 + 90 = 497; 497 : 7 = 71;

#207. 85 x 3 = 255; 255 (1/5 of it) = 51; 51 : 17 = 3; 3 + 79 = 82; 82 - 69 = 13; 13 + 96 = 109;

#208. 135 - 78 = 57; 57 x 4 = 228; 228 (1/4 of it) = 57; 57 : 19 = 3; 3 + 2 = 5; 5 + 62 = 67;

#209. 121 : 11 = 11; 11 + 85 = 96; 96 - 56 = 40; 40 (5/8 of it) = 25; 25 x 4 = 100; 100 + 37 = 137;

#210. 111 (4/6 of it) = 74; 74 : 2 = 37; 37 + 14 = 51; 51 - 26 = 25; 25 x 9 = 225; 225 + 6 = 231;

#211. 56 (1/4 of it) = 14; 14 + 8 = 22; 22 x 8 = 176; 176 + 58 = 234; 234 - 15 = 219; 219 : 3 = 73;

#212. 108 - 97 = 11; 11 + 80 = 91; 91 x 3 = 273; 273 (3/7 of it) = 117; 117 : 39 = 3; 3 + 24 = 27;

#213. 119 (2/7 of it) = 34; 34 x 8 = 272; 272 - 78 = 194; 194 : 2 = 97; 97 + 75 = 172; 172 - 79 = 93;

#214. 126 - 91 = 35; 35 : 5 = 7; 7 + 65 = 72; 72 + 98 = 170; 170 (1/5 of it) = 34; 34 x 6 = 204;

#215. 141 + 89 = 230; 230 : 10 = 23; 23 + 95 = 118; 118 - 94 = 24; 24 (6/8 of it) = 18; 18 x 7 = 126;

#216. 27 + 75 = 102; 102 : 3 = 34; 34 x 5 = 170; 170 - 70 = 100; 100 (1/5 of it) = 20; 20 + 95 = 115;

#217. 92 : 4 = 23; 23 + 36 = 59; 59 + 3 = 62; 62 x 4 = 248; 248 (6/8 of it) = 186; 186 - 50 = 136;

#218. 137 + 16 = 153; 153 - 58 = 95; 95 : 19 = 5; 5 + 30 = 35; 35 (2/7 of it) = 10; 10 x 3 = 30;

#219. 85 - 19 = 66; 66 x 5 = 330; 330 (4/6 of it) = 220; 220 - 11 = 209; 209 : 19 = 11; 11 + 32 = 43;

#220. 144 : 8 = 18; 18 + 51 = 69; 69 - 40 = 29; 29 + 81 = 110; 110 (4/5 of it) = 88; 88 x 4 = 352;

#221. 45 x 9 = 405; 405 (2/5 of it) = 162; 162 - 44 = 118; 118 : 2 = 59; 59 + 97 = 156; 156 - 55 = 101;

#222. 93 (4/6 of it) = 62; 62 - 36 = 26; 26 x 6 = 156; 156 + 50 = 206; 206 : 2 = 103; 103 + 79 = 182;

#223. 54 : 6 = 9; 9 (square it) = 81; 81 (2/3 of it) = 54; 54 - 27 = 27; 27 x 9 = 243; 243 - 42 = 201;

#224. 98 x 3 = 294; 294 (5/6 of it) = 245; 245 : 35 = 7; 7 + 16 = 23; 23 + 69 = 92; 92 - 2 = 90;

#225. 126 + 30 = 156; 156 : 4 = 39; 39 + 6 = 45; 45 (3/5 of it) = 27; 27 x 4 = 108; 108 - 4 = 104;

#226. 75 : 25 = 3; 3 + 60 = 63; 63 (2/7 of it) = 18; 18 x 5 = 90; 90 - 78 = 12; 12 + 8 = 20;

#227. 129 + 19 = 148; 148 - 66 = 82; 82 : 41 = 2; 2 + 48 = 50; 50 x 4 = 200; 200 (2/5 of it) = 80;

#228. 70 x 7 = 490; 490 (4/5 of it) = 392; 392 - 51 = 341; 341 + 62 = 403; 403 : 13 = 31; 31 + 7 = 38;

#229. 77 (5/7 of it) = 55; 55 x 6 = 330; 330 - 27 = 303; 303 - 16 = 287; 287 : 41 = 7; 7 + 86 = 93;

#230. 44 + 48 = 92; 92 x 9 = 828; 828 (1/4 of it) = 207; 207 - 47 = 160; 160 : 10 = 16; 16 + 12 = 28;

#231. 53 + 99 = 152; 152 - 68 = 84; 84 (4/6 of it) = 56; 56 : 4 = 14; 14 x 9 = 126; 126 - 83 = 43;

#232. 115 - 67 = 48; 48 : 8 = 6; 6 + 65 = 71; 71 + 55 = 126; 126 (4/7 of it) = 72; 72 x 6 = 432;

#233. 30 x 7 = 210; 210 - 94 = 116; 116 (2/8 of it) = 29; 29 + 87 = 116; 116 - 2 = 114; 114 : 57 = 2;

#234. 147 (2/7 of it) = 42; 42 x 5 = 210; 210 - 64 = 146; 146 - 49 = 97; 97 + 25 = 122; 122 : 2 = 61;

#235. 35 : 5 = 7; 7 + 81 = 88; 88 x 7 = 616; 616 (7/8 of it) = 539; 539 - 91 = 448; 448 + 30 = 478;

#236. 25 + 81 = 106; 106 : 53 = 2; 2 + 70 = 72; 72 - 27 = 45; 45 (3/5 of it) = 27; 27 x 8 = 216;

#237. 36 x 3 = 108; 108 + 61 = 169; 169 - 73 = 96; 96 : 6 = 16; 16 + 94 = 110; 110 (1/5 of it) = 22;

#238. 140 + 26 = 166; 166 - 62 = 104; 104 : 8 = 13; 13 + 91 = 104; 104 (5/8 of it) = 65; 65 x 4 = 260;

#239. 70 x 3 = 210; 210 + 74 = 284; 284 - 8 = 276; 276 : 3 = 92; 92 + 60 = 152; 152 (5/8 of it) = 95;

#240. 24 (1/6 of it) = 4; 4 x 3 = 12; 12 (square it) = 144; 144 + 68 = 212; 212 : 4 = 53; 53 + 36 = 89;

#241. 105 - 44 = 61; 61 + 56 = 117; 117 : 3 = 39; 39 x 3 = 117; 117 + 75 = 192; 192 (7/8 of it) = 168;

#242. 113 + 70 = 183; 183 - 82 = 101; 101 + 59 = 160; 160 (2/8 of it) = 40; 40 : 10 = 4; 4 x 4 = 16;

#243. 116 (6/8 of it) = 87; 87 - 53 = 34; 34 + 82 = 116; 116 : 29 = 4; 4 (square it) = 16; 16 x 3 = 48;

#244. 43 + 11 = 54; 54 x 3 = 162; 162 - 74 = 88; 88 (3/8 of it) = 33; 33 : 11 = 3; 3 + 96 = 99;

#245. 134 - 77 = 57; 57 x 8 = 456; 456 (1/8 of it) = 57; 57 - 21 = 36; 36 + 26 = 62; 62 : 2 = 31;

#246. 75 - 34 = 41; 41 + 69 = 110; 110 : 55 = 2; 2 + 19 = 21; 21 x 3 = 63; 63 (1/7 of it) = 9;

#247. 139 + 24 = 163; 163 + 73 = 236; 236 : 2 = 118; 118 - 73 = 45; 45 x 3 = 135; 135 (2/3 of it) = 90;

#248. 47 + 53 = 100; 100 (4/5 of it) = 80; 80 : 4 = 20; 20 + 6 = 26; 26 x 5 = 130; 130 - 71 = 59;

#249. 114 : 6 = 19; 19 + 3 = 22; 22 x 8 = 176; 176 - 24 = 152; 152 (5/8 of it) = 95; 95 - 84 = 11;

#250. 97 + 24 = 121; 121 : 11 = 11; 11 + 51 = 62; 62 x 6 = 372; 372 (5/6 of it) = 310; 310 - 81 = 229;

#251. 58 + 18 = 76; 76 x 6 = 456; 456 (50% of it) = 228; 228 + 4 = 232; 232 - 50 = 182; 182 : 7 = 26;

#252. 95 x 9 = 855; 855 + 92 = 947; 947 + 14 = 961; 961 - 46 = 915; 915 : 3 = 305; 305 (80% of it) = 244;

#253. 49 x 9 = 441; 441 - 45 = 396; 396 - 37 = 359; 359 + 36 = 395; 395 (40% of it) = 158; 158 : 2 = 79;

#254. 111 - 33 = 78; 78 - 24 = 54; 54 (50% of it) = 27; 27 x 5 = 135; 135 : 5 = 27; 27 + 23 = 50;

#255. 131 + 24 = 155; 155 - 85 = 70; 70 (80% of it) = 56; 56 x 3 = 168; 168 : 24 = 7; 7 + 22 = 29;

#256. 90 (70% of it) = 63; 63 - 19 = 44; 44 x 7 = 308; 308 + 37 = 345; 345 : 23 = 15; 15 (square it) = 225;

#257. 70 (30% of it) = 21; 21 x 4 = 84; 84 : 14 = 6; 6 + 61 = 67; 67 + 99 = 166; 166 - 63 = 103;

#258. 110 (70% of it) = 77; 77 x 7 = 539; 539 + 27 = 566; 566 + 36 = 602; 602 : 2 = 301; 301 - 69 = 232;

#259. 69 : 23 = 3; 3 + 96 = 99; 99 - 23 = 76; 76 + 24 = 100; 100 (20% of it) = 20; 20 x 4 = 80;

#260. 85 : 5 = 17; 17 + 73 = 90; 90 x 4 = 360; 360 (40% of it) = 144; 144 - 32 = 112; 112 - 75 = 37;

#261. 110 (60% of it) = 66; 66 : 2 = 33; 33 x 4 = 132; 132 + 82 = 214; 214 - 45 = 169; 169 + 17 = 186;

#262. 106 - 54 = 52; 52 : 26 = 2; 2 + 56 = 58; 58 - 24 = 34; 34 x 5 = 170; 170 (50% of it) = 85;

#263. 60 - 20 = 40; 40 (40% of it) = 16; 16 + 89 = 105; 105 : 5 = 21; 21 x 4 = 84; 84 + 87 = 171;

#264. 117 : 13 = 9; 9 x 4 = 36; 36 + 18 = 54; 54 (50% of it) = 27; 27 + 51 = 78; 78 - 4 = 74;

#265. 100 + 82 = 182; 182 (50% of it) = 91; 91 + 4 = 95; 95 - 61 = 34; 34 x 4 = 136; 136 : 4 = 34;

#266. 74 - 38 = 36; 36 (50% of it) = 18; 18 + 69 = 87; 87 x 4 = 348; 348 : 12 = 29; 29 + 16 = 45;

#267. 98 x 9 = 882; 882 - 33 = 849; 849 : 3 = 283; 283 + 99 = 382; 382 (50% of it) = 191; 191 + 3 = 194;

#268. 24 (50% of it) = 12; 12 + 4 = 16; 16 x 7 = 112; 112 - 2 = 110; 110 + 26 = 136; 136 : 4 = 34;

#269. 125 : 5 = 25; 25 (80% of it) = 20; 20 + 6 = 26; 26 x 7 = 182; 182 - 93 = 89; 89 + 79 = 168;

#270. 73 + 23 = 96; 96 : 4 = 24; 24 x 6 = 144; 144 (50% of it) = 72; 72 - 41 = 31; 31 + 27 = 58;

#271. 143 - 98 = 45; 45 (80% of it) = 36; 36 x 4 = 144; 144 + 69 = 213; 213 : 3 = 71; 71 + 17 = 88;

#272. 94 x 7 = 658; 658 - 6 = 652; 652 + 18 = 670; 670 (80% of it) = 536; 536 + 96 = 632; 632 : 4 = 158;

#273. 45 (20% of it) = 9; 9 x 5 = 45; 45 : 3 = 15; 15 + 97 = 112; 112 + 50 = 162; 162 - 76 = 86;

#274. 105 (40% of it) = 42; 42 x 5 = 210; 210 + 83 = 293; 293 + 16 = 309; 309 - 4 = 305; 305 : 5 = 61;

#275. 71 + 67 = 138; 138 - 68 = 70; 70 (40% of it) = 28; 28 x 5 = 140; 140 : 7 = 20; 20 + 5 = 25;

#276. 94 - 24 = 70; 70 : 10 = 7; 7 + 8 = 15; 15 x 9 = 135; 135 (40% of it) = 54; 54 + 92 = 146;

#277. 141 : 47 = 3; 3 + 7 = 10; 10 (square it) = 100; 100 - 55 = 45; 45 x 9 = 405; 405 (20% of it) = 81;

#278. 78 - 13 = 65; 65 (40% of it) = 26; 26 x 8 = 208; 208 : 26 = 8; 8 (square it) = 64; 64 - 50 = 14;

#279. 93 + 29 = 122; 122 + 34 = 156; 156 (50% of it) = 78; 78 x 7 = 546; 546 - 7 = 539; 539 : 7 = 77;

#280. 82 - 66 = 16; 16 x 3 = 48; 48 : 8 = 6; 6 + 18 = 24; 24 + 61 = 85; 85 (80% of it) = 68;

#281. 100 : 2 = 50; 50 (30% of it) = 15; 15 + 26 = 41; 41 + 3 = 44; 44 x 7 = 308; 308 - 74 = 234;

#282. 75 x 7 = 525; 525 : 35 = 15; 15 (square it) = 225; 225 - 36 = 189; 189 + 7 = 196; 196 (50% of it) = 98;

#283. 60 - 25 = 35; 35 (40% of it) = 14; 14 x 8 = 112; 112 + 64 = 176; 176 - 7 = 169; 169 : 13 = 13;

#284. 36 : 18 = 2; 2 + 24 = 26; 26 + 69 = 95; 95 - 43 = 52; 52 x 5 = 260; 260 (30% of it) = 78;

#285. 31 + 73 = 104; 104 - 49 = 55; 55 (80% of it) = 44; 44 x 7 = 308; 308 - 16 = 292; 292 : 2 = 146;

#286. 85 x 6 = 510; 510 (30% of it) = 153; 153 - 96 = 57; 57 : 19 = 3; 3 + 16 = 19; 19 + 72 = 91;

#287. 125 + 48 = 173; 173 + 63 = 236; 236 (50% of it) = 118; 118 - 60 = 58; 58 x 8 = 464; 464 : 16 = 29;

#288. 108 - 49 = 59; 59 + 58 = 117; 117 - 85 = 32; 32 : 4 = 8; 8 x 5 = 40; 40 (80% of it) = 32;

#289. 54 - 26 = 28; 28 (50% of it) = 14; 14 + 59 = 73; 73 + 35 = 108; 108 : 27 = 4; 4 x 5 = 20;

#290. 61 + 94 = 155; 155 (60% of it) = 93; 93 x 7 = 651; 651 - 15 = 636; 636 : 6 = 106; 106 - 87 = 19;

#291. 62 + 23 = 85; 85 x 6 = 510; 510 - 92 = 418; 418 : 22 = 19; 19 + 6 = 25; 25 (80% of it) = 20;

#292. 132 - 81 = 51; 51 + 60 = 111; 111 - 96 = 15; 15 x 8 = 120; 120 : 2 = 60; 60 (30% of it) = 18;

#293. 145 (40% of it) = 58; 58 x 3 = 174; 174 : 29 = 6; 6 + 51 = 57; 57 - 24 = 33; 33 + 2 = 35;

#294. 62 + 90 = 152; 152 : 19 = 8; 8 x 8 = 64; 64 - 32 = 32; 32 + 92 = 124; 124 (50% of it) = 62;

#295. 98 x 6 = 588; 588 (50% of it) = 294; 294 - 7 = 287; 287 : 7 = 41; 41 + 69 = 110; 110 - 17 = 93;

#296. 74 : 37 = 2; 2 + 88 = 90; 90 x 4 = 360; 360 + 95 = 455; 455 (80% of it) = 364; 364 - 63 = 301;

#297. 56 - 17 = 39; 39 x 7 = 273; 273 + 66 = 339; 339 : 3 = 113; 113 + 7 = 120; 120 (40% of it) = 48;

#298. 53 + 77 = 130; 130 (40% of it) = 52; 52 x 7 = 364; 364 - 60 = 304; 304 - 22 = 282; 282 : 47 = 6;

#299. 100 : 5 = 20; 20 x 8 = 160; 160 (30% of it) = 48; 48 + 63 = 111; 111 + 77 = 188; 188 - 6 = 182;

#300. 58 - 26 = 32; 32 : 2 = 16; 16 x 7 = 112; 112 (50% of it) = 56; 56 - 39 = 17; 17 + 89 = 106;

#301. 94 - 78 = 16; 16 x 5 = 80; 80 (10% of it) = 8; 8 (square it) = 64; 64 (1/4 of it) = 16; 16 + 80 = 96;

#302. 75 (4/6 of it) = 50; 50 x 8 = 400; 400 + 45 = 445; 445 (60% of it) = 267; 267 - 64 = 203; 203 : 7 = 29;

#303. 100 : 25 = 4; 4 + 61 = 65; 65 (40% of it) = 26; 26 x 7 = 182; 182 - 32 = 150; 150 (4/6 of it) = 100;

#304. 59 + 70 = 129; 129 - 79 = 50; 50 (20% of it) = 10; 10 x 7 = 70; 70 (4/5 of it) = 56; 56 : 7 = 8;

#305. 52 (50% of it) = 26; 26 x 7 = 182; 182 - 78 = 104; 104 (3/8 of it) = 39; 39 : 3 = 13; 13 + 55 = 68;

#306. 74 x 6 = 444; 444 : 3 = 148; 148 - 73 = 75; 75 (4/6 of it) = 50; 50 (10% of it) = 5; 5 + 72 = 77;

#307. 82 (50% of it) = 41; 41 + 69 = 110; 110 (2/5 of it) = 44; 44 : 11 = 4; 4 (square it) = 16; 16 x 5 = 80;

#308. 82 x 5 = 410; 410 : 5 = 82; 82 (50% of it) = 41; 41 + 95 = 136; 136 (7/8 of it) = 119; 119 - 35 = 84;

#309. 76 x 7 = 532; 532 (3/7 of it) = 228; 228 - 46 = 182; 182 + 93 = 275; 275 : 5 = 55; 55 (60% of it) = 33;

#310. 100 (50% of it) = 50; 50 + 69 = 119; 119 - 47 = 72; 72 : 9 = 8; 8 x 9 = 72; 72 (1/8 of it) = 9;

#311. 82 x 9 = 738; 738 (50% of it) = 369; 369 - 44 = 325; 325 + 95 = 420; 420 : 6 = 70; 70 (5/7 of it) = 50;

#312. 102 (5/6 of it) = 85; 85 (20% of it) = 17; 17 + 67 = 84; 84 : 21 = 4; 4 (square it) = 16; 16 x 3 = 48;

#313. 65 (80% of it) = 52; 52 - 25 = 27; 27 (2/3 of it) = 18; 18 x 7 = 126; 126 : 7 = 18; 18 + 60 = 78;

#314. 102 : 2 = 51; 51 x 3 = 153; 153 - 104 = 49; 49 (1/7 of it) = 7; 7 + 78 = 85; 85 (80% of it) = 68;

#315. 80 : 5 = 16; 16 x 9 = 144; 144 (50% of it) = 72; 72 - 60 = 12; 12 (square it) = 144; 144 (3/8 of it) = 54;

#316. 70 (4/7 of it) = 40; 40 : 10 = 4; 4 + 74 = 78; 78 - 64 = 14; 14 x 5 = 70; 70 (50% of it) = 35;

#317. 58 - 30 = 28; 28 (1/4 of it) = 7; 7 + 44 = 51; 51 x 5 = 255; 255 (20% of it) = 51; 51 : 3 = 17;

#318. 99 + 69 = 168; 168 (6/7 of it) = 144; 144 (50% of it) = 72; 72 : 9 = 8; 8 (square it) = 64; 64 x 9 = 576;

#319. 59 + 66 = 125; 125 - 89 = 36; 36 (50% of it) = 18; 18 x 4 = 72; 72 : 3 = 24; 24 (1/6 of it) = 4;

#320. 107 + 81 = 188; 188 (50% of it) = 94; 94 x 4 = 376; 376 - 72 = 304; 304 (6/8 of it) = 228; 228 : 3 = 76;

#321. 71 + 90 = 161; 161 - 93 = 68; 68 (50% of it) = 34; 34 x 4 = 136; 136 (1/4 of it) = 34; 34 : 17 = 2;

#322. 92 - 53 = 39; 39 : 13 = 3; 3 + 87 = 90; 90 x 4 = 360; 360 (1/6 of it) = 60; 60 (70% of it) = 42;

#323. 74 x 5 = 370; 370 (70% of it) = 259; 259 : 37 = 7; 7 + 109 = 116; 116 - 86 = 30; 30 (5/6 of it) = 25;

#324. 97 + 62 = 159; 159 - 79 = 80; 80 (3/5 of it) = 48; 48 : 2 = 24; 24 (50% of it) = 12; 12 (square it) = 144;

#325. 108 (6/8 of it) = 81; 81 x 6 = 486; 486 (50% of it) = 243; 243 - 78 = 165; 165 : 55 = 3; 3 + 77 = 80;

#326. 94 - 45 = 49; 49 (2/7 of it) = 14; 14 x 5 = 70; 70 : 14 = 5; 5 + 90 = 95; 95 (80% of it) = 76;

#327. 112 : 28 = 4; 4 + 70 = 74; 74 x 3 = 222; 222 - 62 = 160; 160 (60% of it) = 96; 96 (3/8 of it) = 36;

#328. 81 - 36 = 45; 45 : 15 = 3; 3 + 69 = 72; 72 (3/4 of it) = 54; 54 (50% of it) = 27; 27 x 5 = 135;

#329. 82 (50% of it) = 41; 41 + 51 = 92; 92 (6/8 of it) = 69; 69 x 3 = 207; 207 - 53 = 154; 154 : 22 = 7;

#330. 101 + 88 = 189; 189 - 112 = 77; 77 (6/7 of it) = 66; 66 x 5 = 330; 330 : 3 = 110; 110 (30% of it) = 33;

#331. 113 + 48 = 161; 161 (6/7 of it) = 138; 138 (50% of it) = 69; 69 x 4 = 276; 276 : 23 = 12; 12 (square it) = 144;

#332. 110 (50% of it) = 55; 55 : 11 = 5; 5 + 60 = 65; 65 x 5 = 325; 325 (2/5 of it) = 130; 130 - 112 = 18;

#333. 101 + 70 = 171; 171 - 81 = 90; 90 (3/5 of it) = 54; 54 x 3 = 162; 162 (50% of it) = 81; 81 : 27 = 3;

#334. 107 + 67 = 174; 174 (50% of it) = 87; 87 x 5 = 435; 435 - 95 = 340; 340 (1/4 of it) = 85; 85 : 17 = 5;

#335. 100 (10% of it) = 10; 10 x 9 = 90; 90 (1/6 of it) = 15; 15 (square it) = 225; 225 - 66 = 159; 159 : 53 = 3;

#336. 108 (50% of it) = 54; 54 - 35 = 19; 19 + 97 = 116; 116 : 29 = 4; 4 x 6 = 24; 24 (2/3 of it) = 16;

#337. 113 + 69 = 182; 182 - 106 = 76; 76 x 3 = 228; 228 (2/3 of it) = 152; 152 : 4 = 38; 38 (50% of it) = 19;

#338. 76 : 38 = 2; 2 + 108 = 110; 110 (80% of it) = 88; 88 (2/8 of it) = 22; 22 x 7 = 154; 154 - 63 = 91;

#339. 91 (6/7 of it) = 78; 78 x 3 = 234; 234 + 76 = 310; 310 (40% of it) = 124; 124 - 109 = 15; 15 (square it) = 225;

#340. 109 + 101 = 210; 210 (10% of it) = 21; 21 (2/3 of it) = 14; 14 (square it) = 196; 196 : 4 = 49; 49 x 9 = 441;

#341. 105 (2/7 of it) = 30; 30 (30% of it) = 9; 9 x 9 = 81; 81 - 62 = 19; 19 + 50 = 69; 69 : 3 = 23;

#342. 101 + 99 = 200; 200 - 86 = 114; 114 (2/3 of it) = 76; 76 (50% of it) = 38; 38 x 7 = 266; 266 : 2 = 133;

#343. 79 + 107 = 186; 186 (5/6 of it) = 155; 155 (40% of it) = 62; 62 x 7 = 434; 434 : 2 = 217; 217 - 75 = 142;

#344. 94 x 9 = 846; 846 (5/6 of it) = 705; 705 - 115 = 590; 590 (20% of it) = 118; 118 : 2 = 59; 59 + 118 = 177;

#345. 89 + 111 = 200; 200 (60% of it) = 120; 120 : 2 = 60; 60 - 46 = 14; 14 x 3 = 42; 42 (4/6 of it) = 28;

#346. 76 + 92 = 168; 168 - 113 = 55; 55 x 8 = 440; 440 (40% of it) = 176; 176 (2/8 of it) = 44; 44 : 11 = 4;

#347. 101 + 97 = 198; 198 - 78 = 120; 120 (1/8 of it) = 15; 15 x 8 = 120; 120 (10% of it) = 12; 12 (square it) = 144;

#348. 108 - 66 = 42; 42 (50% of it) = 21; 21 (6/7 of it) = 18; 18 + 122 = 140; 140 : 2 = 70; 70 x 6 = 420;

#349. 92 (50% of it) = 46; 46 x 7 = 322; 322 (5/7 of it) = 230; 230 : 23 = 10; 10 + 116 = 126; 126 - 78 = 48;

#350. 113 + 94 = 207; 207 : 3 = 69; 69 x 8 = 552; 552 (2/3 of it) = 368; 368 - 103 = 265; 265 (80% of it) = 212;

#351. 98 - 59 = 39; 39 x 5 = 195; 195 (2/3 of it) = 130; 130 (30% of it) = 39; 39 : 3 = 13; 13 + 76 = 89;

#352. 89 + 114 = 203; 203 - 95 = 108; 108 : 9 = 12; 12 x 10 = 120; 120 (7/8 of it) = 105; 105 (20% of it) = 21;

#353. 102 : 34 = 3; 3 + 112 = 115; 115 (80% of it) = 92; 92 x 10 = 920; 920 (1/4 of it) = 230; 230 - 94 = 136;

#354. 99 : 11 = 9; 9 + 102 = 111; 111 - 75 = 36; 36 (50% of it) = 18; 18 x 6 = 108; 108 (5/6 of it) = 90;

#355. 107 + 70 = 177; 177 - 85 = 92; 92 x 8 = 736; 736 (5/8 of it) = 460; 460 : 10 = 46; 46 (50% of it) = 23;

#356. 126 (2/7 of it) = 36; 36 (50% of it) = 18; 18 x 3 = 54; 54 - 21 = 33; 33 : 11 = 3; 3 + 116 = 119;

#357. 94 - 48 = 46; 46 : 2 = 23; 23 + 69 = 92; 92 (3/4 of it) = 69; 69 x 8 = 552; 552 (50% of it) = 276;

#358. 111 (2/3 of it) = 74; 74 x 10 = 740; 740 (70% of it) = 518; 518 + 121 = 639; 639 - 60 = 579; 579 : 3 = 193;

#359. 126 (50% of it) = 63; 63 - 25 = 38; 38 x 3 = 114; 114 (4/6 of it) = 76; 76 : 2 = 38; 38 + 106 = 144;

#360. 107 + 64 = 171; 171 - 123 = 48; 48 (2/3 of it) = 32; 32 (50% of it) = 16; 16 x 8 = 128; 128 : 16 = 8;

#361. 90 (1/5 of it) = 18; 18 + 67 = 85; 85 - 65 = 20; 20 x 4 = 80; 80 (40% of it) = 32; 32 : 16 = 2;

#362. 109 + 66 = 175; 175 (5/7 of it) = 125; 125 (20% of it) = 25; 25 x 9 = 225; 225 - 83 = 142; 142 : 71 = 2;

#363. 129 (2/3 of it) = 86; 86 x 3 = 258; 258 - 124 = 134; 134 (50% of it) = 67; 67 + 76 = 143; 143 : 11 = 13;

#364. 130 (20% of it) = 26; 26 + 61 = 87; 87 x 10 = 870; 870 (1/5 of it) = 174; 174 - 74 = 100; 100 : 20 = 5;

#365. 105 : 7 = 15; 15 (square it) = 225; 225 - 124 = 101; 101 + 119 = 220; 220 (6/8 of it) = 165; 165 (20% of it) = 33;

#366. 111 - 84 = 27; 27 x 8 = 216; 216 (5/8 of it) = 135; 135 (80% of it) = 108; 108 : 12 = 9; 9 + 106 = 115;

#367. 112 (50% of it) = 56; 56 (4/7 of it) = 32; 32 x 3 = 96; 96 : 8 = 12; 12 (square it) = 144; 144 - 95 = 49;

#368. 111 : 3 = 37; 37 + 95 = 132; 132 (3/4 of it) = 99; 99 - 71 = 28; 28 x 10 = 280; 280 (40% of it) = 112;

#369. 127 + 86 = 213; 213 - 78 = 135; 135 : 15 = 9; 9 x 9 = 81; 81 (4/6 of it) = 54; 54 (50% of it) = 27;

#370. 90 (5/6 of it) = 75; 75 - 24 = 51; 51 x 10 = 510; 510 (20% of it) = 102; 102 : 51 = 2; 2 + 74 = 76;

#371. 93 - 59 = 34; 34 x 5 = 170; 170 : 2 = 85; 85 (20% of it) = 17; 17 + 109 = 126; 126 (5/7 of it) = 90;

#372. 103 + 65 = 168; 168 - 99 = 69; 69 x 6 = 414; 414 : 2 = 207; 207 (4/6 of it) = 138; 138 (50% of it) = 69;

#373. 133 (5/7 of it) = 95; 95 x 7 = 665; 665 (20% of it) = 133; 133 + 93 = 226; 226 - 131 = 95; 95 : 19 = 5;

#374. 111 : 3 = 37; 37 + 95 = 132; 132 (6/8 of it) = 99; 99 x 4 = 396; 396 (50% of it) = 198; 198 - 123 = 75;

#375. 100 (1/4 of it) = 25; 25 (80% of it) = 20; 20 x 4 = 80; 80 : 8 = 10; 10 (square it) = 100; 100 - 32 = 68;

#376. 87 : 29 = 3; 3 + 67 = 70; 70 (40% of it) = 28; 28 (5/7 of it) = 20; 20 x 5 = 100; 100 - 65 = 35;

#377. 134 (50% of it) = 67; 67 + 115 = 182; 182 (4/7 of it) = 104; 104 : 26 = 4; 4 (square it) = 16; 16 x 8 = 128;

#378. 100 (6/8 of it) = 75; 75 (80% of it) = 60; 60 : 6 = 10; 10 + 68 = 78; 78 x 7 = 546; 546 - 96 = 450;

#379. 95 x 5 = 475; 475 : 19 = 25; 25 (20% of it) = 5; 5 + 129 = 134; 134 - 20 = 114; 114 (4/6 of it) = 76;

#380. 129 - 97 = 32; 32 x 10 = 320; 320 (10% of it) = 32; 32 (3/8 of it) = 12; 12 (square it) = 144; 144 : 18 = 8;

#381. 99 - 73 = 26; 26 x 5 = 130; 130 : 26 = 5; 5 + 115 = 120; 120 (30% of it) = 36; 36 (2/3 of it) = 24;

#382. 95 (80% of it) = 76; 76 : 4 = 19; 19 + 123 = 142; 142 - 84 = 58; 58 x 8 = 464; 464 (1/4 of it) = 116;

#383. 109 + 119 = 228; 228 - 120 = 108; 108 : 9 = 12; 12 x 10 = 120; 120 (80% of it) = 96; 96 (4/6 of it) = 64;

#384. 95 (2/5 of it) = 38; 38 + 133 = 171; 171 - 87 = 84; 84 (50% of it) = 42; 42 : 7 = 6; 6 (square it) = 36;

#385. 116 + 96 = 212; 212 - 87 = 125; 125 : 5 = 25; 25 (40% of it) = 10; 10 x 8 = 80; 80 (3/8 of it) = 30;

#386. 124 : 62 = 2; 2 + 114 = 116; 116 - 72 = 44; 44 x 4 = 176; 176 (3/4 of it) = 132; 132 (50% of it) = 66;

#387. 125 : 25 = 5; 5 + 100 = 105; 105 (80% of it) = 84; 84 - 62 = 22; 22 x 8 = 176; 176 (6/8 of it) = 132;

#388. 134 - 91 = 43; 43 + 107 = 150; 150 (30% of it) = 45; 45 : 3 = 15; 15 x 4 = 60; 60 (2/8 of it) = 15;

#389. 97 + 140 = 237; 237 - 123 = 114; 114 : 3 = 38; 38 x 5 = 190; 190 (50% of it) = 95; 95 (4/5 of it) = 76;

#390. 102 (50% of it) = 51; 51 : 3 = 17; 17 + 137 = 154; 154 (1/7 of it) = 22; 22 x 3 = 66; 66 - 27 = 39;

#391. 137 + 124 = 261; 261 : 29 = 9; 9 (square it) = 81; 81 x 5 = 405; 405 (2/3 of it) = 270; 270 (60% of it) = 162;

#392. 125 (1/5 of it) = 25; 25 x 9 = 225; 225 - 133 = 92; 92 : 2 = 46; 46 (50% of it) = 23; 23 + 99 = 122;

#393. 126 - 84 = 42; 42 (5/7 of it) = 30; 30 (40% of it) = 12; 12 + 140 = 152; 152 : 2 = 76; 76 x 4 = 304;

#394. 102 (4/6 of it) = 68; 68 : 4 = 17; 17 + 128 = 145; 145 (60% of it) = 87; 87 x 9 = 783; 783 - 128 = 655;

#395. 136 : 8 = 17; 17 + 88 = 105; 105 (3/7 of it) = 45; 45 (40% of it) = 18; 18 x 4 = 72; 72 - 23 = 49;

#396. 110 (60% of it) = 66; 66 : 3 = 22; 22 x 10 = 220; 220 (1/4 of it) = 55; 55 - 38 = 17; 17 + 135 = 152;

#397. 124 : 62 = 2; 2 + 85 = 87; 87 - 62 = 25; 25 (80% of it) = 20; 20 x 5 = 100; 100 (6/8 of it) = 75;

#398. 109 + 112 = 221; 221 - 146 = 75; 75 : 5 = 15; 15 (square it) = 225; 225 (4/5 of it) = 180; 180 (20% of it) = 36;

#399. 132 + 113 = 245; 245 (1/7 of it) = 35; 35 (60% of it) = 21; 21 x 9 = 189; 189 : 21 = 9; 9 (square it) = 81;

#400. 137 + 148 = 285; 285 (80% of it) = 228; 228 (6/8 of it) = 171; 171 : 3 = 57; 57 x 3 = 171; 171 - 146 = 25;

#401. 102 : 51 = 2; 2 + 140 = 142; 142 - 115 = 27; 27 x 10 = 270; 270 (1/6 of it) = 45; 45 (80% of it) = 36;

#402. 133 - 51 = 82; 82 : 41 = 2; 2 + 118 = 120; 120 (2/8 of it) = 30; 30 (30% of it) = 9; 9 x 10 = 90;

#403. 110 (40% of it) = 44; 44 : 2 = 22; 22 x 4 = 88; 88 - 36 = 52; 52 (2/8 of it) = 13; 13 + 108 = 121;

#404. 141 (2/3 of it) = 94; 94 x 11 = 1034; 1034 - 142 = 892; 892 : 2 = 446; 446 + 89 = 535; 535 (60% of it) = 321;

#405. 144 : 18 = 8; 8 x 6 = 48; 48 (5/8 of it) = 30; 30 (10% of it) = 3; 3 + 130 = 133; 133 - 92 = 41;

#406. 141 : 3 = 47; 47 + 117 = 164; 164 - 116 = 48; 48 (5/8 of it) = 30; 30 (60% of it) = 18; 18 x 6 = 108;

#407. 128 - 46 = 82; 82 : 2 = 41; 41 + 109 = 150; 150 (4/6 of it) = 100; 100 (10% of it) = 10; 10 x 3 = 30;

#408. 114 - 94 = 20; 20 + 92 = 112; 112 (7/8 of it) = 98; 98 : 2 = 49; 49 x 8 = 392; 392 (50% of it) = 196;

#409. 144 (2/8 of it) = 36; 36 x 11 = 396; 396 : 33 = 12; 12 + 153 = 165; 165 - 110 = 55; 55 (20% of it) = 11;

#410. 126 (4/6 of it) = 84; 84 : 14 = 6; 6 (square it) = 36; 36 x 10 = 360; 360 (60% of it) = 216; 216 + 95 = 311;

#411. 116 - 84 = 32; 32 : 8 = 4; 4 (square it) = 16; 16 + 136 = 152; 152 (50% of it) = 76; 76 (6/8 of it) = 57;

#412. 126 + 129 = 255; 255 (40% of it) = 102; 102 : 2 = 51; 51 - 36 = 15; 15 x 7 = 105; 105 (3/5 of it) = 63;

#413. 135 : 9 = 15; 15 x 4 = 60; 60 (40% of it) = 24; 24 (1/6 of it) = 4; 4 + 101 = 105; 105 - 27 = 78;

#414. 115 : 5 = 23; 23 + 142 = 165; 165 (60% of it) = 99; 99 - 75 = 24; 24 (2/8 of it) = 6; 6 (square it) = 36;

#415. 126 (2/7 of it) = 36; 36 (50% of it) = 18; 18 x 6 = 108; 108 - 20 = 88; 88 : 44 = 2; 2 + 96 = 98;

#416. 127 + 105 = 232; 232 : 2 = 116; 116 (50% of it) = 58; 58 x 9 = 522; 522 (4/6 of it) = 348; 348 - 113 = 235;

#417. 121 : 11 = 11; 11 + 144 = 155; 155 (40% of it) = 62; 62 x 4 = 248; 248 (6/8 of it) = 186; 186 - 147 = 39;

#418. 147 : 7 = 21; 21 (5/7 of it) = 15; 15 + 100 = 115; 115 (80% of it) = 92; 92 x 5 = 460; 460 - 147 = 313;

#419. 115 : 23 = 5; 5 + 130 = 135; 135 (80% of it) = 108; 108 - 87 = 21; 21 (6/7 of it) = 18; 18 x 7 = 126;

#420. 124 (3/4 of it) = 93; 93 - 44 = 49; 49 x 5 = 245; 245 (40% of it) = 98; 98 : 49 = 2; 2 + 133 = 135;

#421. 149 + 126 = 275; 275 (20% of it) = 55; 55 - 37 = 18; 18 x 7 = 126; 126 (6/7 of it) = 108; 108 : 12 = 9;

#422. 112 - 90 = 22; 22 x 7 = 154; 154 : 77 = 2; 2 + 103 = 105; 105 (5/7 of it) = 75; 75 (20% of it) = 15;

#423. 158 : 79 = 2; 2 + 122 = 124; 124 - 100 = 24; 24 x 6 = 144; 144 (5/6 of it) = 120; 120 (20% of it) = 24;

#424. 120 (70% of it) = 84; 84 x 7 = 588; 588 - 102 = 486; 486 + 119 = 605; 605 (4/5 of it) = 484; 484 : 11 = 44;

#425. 145 (40% of it) = 58; 58 : 2 = 29; 29 + 93 = 122; 122 - 104 = 18; 18 x 8 = 144; 144 (1/4 of it) = 36;

#426. 128 (3/4 of it) = 96; 96 - 83 = 13; 13 + 143 = 156; 156 : 26 = 6; 6 x 5 = 30; 30 (50% of it) = 15;

#427. 124 - 67 = 57; 57 (4/6 of it) = 38; 38 : 19 = 2; 2 + 113 = 115; 115 (40% of it) = 46; 46 x 5 = 230;

#428. 150 (10% of it) = 15; 15 x 10 = 150; 150 - 99 = 51; 51 : 3 = 17; 17 + 102 = 119; 119 (1/7 of it) = 17;

#429. 145 : 29 = 5; 5 + 123 = 128; 128 - 58 = 70; 70 (4/7 of it) = 40; 40 (40% of it) = 16; 16 x 9 = 144;

#430. 130 (70% of it) = 91; 91 x 4 = 364; 364 (5/7 of it) = 260; 260 : 65 = 4; 4 (square it) = 16; 16 + 163 = 179;

#431. 160 (1/5 of it) = 32; 32 + 163 = 195; 195 - 115 = 80; 80 : 8 = 10; 10 (square it) = 100; 100 (40% of it) = 40;

#432. 159 : 3 = 53; 53 + 147 = 200; 200 (40% of it) = 80; 80 x 10 = 800; 800 (1/5 of it) = 160; 160 - 132 = 28;

#433. 157 + 138 = 295; 295 (60% of it) = 177; 177 - 113 = 64; 64 (2/8 of it) = 16; 16 x 11 = 176; 176 : 22 = 8;

#434. 118 - 67 = 51; 51 : 17 = 3; 3 + 97 = 100; 100 (4/5 of it) = 80; 80 (80% of it) = 64; 64 x 8 = 512;

#435. 152 (5/8 of it) = 95; 95 (40% of it) = 38; 38 : 2 = 19; 19 + 103 = 122; 122 - 100 = 22; 22 x 11 = 242;

#436. 117 - 68 = 49; 49 x 5 = 245; 245 (60% of it) = 147; 147 (2/7 of it) = 42; 42 + 141 = 183; 183 : 3 = 61;

#437. 134 - 58 = 76; 76 (1/4 of it) = 19; 19 + 145 = 164; 164 : 41 = 4; 4 x 10 = 40; 40 (10% of it) = 4;

#438. 146 : 2 = 73; 73 + 112 = 185; 185 (60% of it) = 111; 111 - 99 = 12; 12 (square it) = 144; 144 (1/4 of it) = 36;

#439. 148 (1/4 of it) = 37; 37 + 153 = 190; 190 (50% of it) = 95; 95 - 27 = 68; 68 x 4 = 272; 272 : 8 = 34;

#440. 143 : 11 = 13; 13 + 141 = 154; 154 - 140 = 14; 14 x 10 = 140; 140 (80% of it) = 112; 112 (5/7 of it) = 80;

#441. 152 (7/8 of it) = 133; 133 - 83 = 50; 50 (30% of it) = 15; 15 x 10 = 150; 150 : 75 = 2; 2 + 154 = 156;

#442. 132 - 108 = 24; 24 (50% of it) = 12; 12 x 4 = 48; 48 : 16 = 3; 3 + 126 = 129; 129 (2/3 of it) = 86;

#443. 138 (50% of it) = 69; 69 - 50 = 19; 19 + 111 = 130; 130 : 5 = 26; 26 x 4 = 104; 104 (3/4 of it) = 78;

#444. 149 + 130 = 279; 279 : 31 = 9; 9 x 11 = 99; 99 - 36 = 63; 63 (5/7 of it) = 45; 45 (40% of it) = 18;

#445. 130 (1/5 of it) = 26; 26 + 119 = 145; 145 (80% of it) = 116; 116 - 68 = 48; 48 : 12 = 4; 4 x 10 = 40;

#446. 160 (1/4 of it) = 40; 40 (30% of it) = 12; 12 x 10 = 120; 120 : 10 = 12; 12 (square it) = 144; 144 + 139 = 283;

#447. 162 : 81 = 2; 2 + 140 = 142; 142 - 52 = 90; 90 (30% of it) = 27; 27 (2/3 of it) = 18; 18 x 7 = 126;

#448. 168 : 28 = 6; 6 x 11 = 66; 66 - 35 = 31; 31 + 145 = 176; 176 (5/8 of it) = 110; 110 (70% of it) = 77;

#449. 163 + 107 = 270; 270 (80% of it) = 216; 216 (3/4 of it) = 162; 162 : 6 = 27; 27 x 5 = 135; 135 - 105 = 30;

#450. 149 + 148 = 297; 297 : 3 = 99; 99 - 29 = 70; 70 (5/7 of it) = 50; 50 (20% of it) = 10; 10 x 8 = 80;

#451. 170 (80% of it) = 136; 136 - 30 = 106; 106 : 53 = 2; 2 + 126 = 128; 128 (1/8 of it) = 16; 16 x 6 = 96;

#452. 143 + 136 = 279; 279 : 3 = 93; 93 x 10 = 930; 930 (40% of it) = 372; 372 (3/4 of it) = 279; 279 - 114 = 165;

#453. 143 : 11 = 13; 13 + 151 = 164; 164 - 134 = 30; 30 (70% of it) = 21; 21 (2/7 of it) = 6; 6 x 11 = 66;

#454. 170 (70% of it) = 119; 119 (2/7 of it) = 34; 34 x 9 = 306; 306 - 149 = 157; 157 + 108 = 265; 265 : 53 = 5;

#455. 136 (6/8 of it) = 102; 102 : 3 = 34; 34 (50% of it) = 17; 17 + 152 = 169; 169 - 153 = 16; 16 x 3 = 48;

#456. 136 (3/4 of it) = 102; 102 - 72 = 30; 30 (30% of it) = 9; 9 (square it) = 81; 81 : 27 = 3; 3 + 161 = 164;

#457. 158 : 2 = 79; 79 + 130 = 209; 209 - 149 = 60; 60 (60% of it) = 36; 36 (3/4 of it) = 27; 27 x 7 = 189;

#458. 162 : 9 = 18; 18 x 7 = 126; 126 + 138 = 264; 264 (6/8 of it) = 198; 198 - 133 = 65; 65 (20% of it) = 13;

#459. 144 + 174 = 318; 318 - 122 = 196; 196 : 14 = 14; 14 x 10 = 140; 140 (70% of it) = 98; 98 (2/7 of it) = 28;

#460. 173 + 147 = 320; 320 - 168 = 152; 152 : 19 = 8; 8 x 5 = 40; 40 (70% of it) = 28; 28 (2/7 of it) = 8;

#461. 154 - 116 = 38; 38 x 10 = 380; 380 (20% of it) = 76; 76 : 19 = 4; 4 + 170 = 174; 174 (5/6 of it) = 145;

#462. 160 (70% of it) = 112; 112 : 28 = 4; 4 (square it) = 16; 16 + 154 = 170; 170 (1/5 of it) = 34; 34 x 6 = 204;

#463. 153 - 76 = 77; 77 : 7 = 11; 11 + 164 = 175; 175 (1/7 of it) = 25; 25 (40% of it) = 10; 10 (square it) = 100;

#464. 152 (2/8 of it) = 38; 38 x 6 = 228; 228 : 4 = 57; 57 - 38 = 19; 19 + 135 = 154; 154 (50% of it) = 77;

#465. 169 : 13 = 13; 13 + 159 = 172; 172 (50% of it) = 86; 86 - 65 = 21; 21 (5/7 of it) = 15; 15 x 12 = 180;

#466. 167 + 127 = 294; 294 : 14 = 21; 21 (5/7 of it) = 15; 15 x 5 = 75; 75 - 25 = 50; 50 (80% of it) = 40;

#467. 133 - 88 = 45; 45 (80% of it) = 36; 36 : 6 = 6; 6 x 12 = 72; 72 (3/8 of it) = 27; 27 + 157 = 184;

#468. 146 - 116 = 30; 30 (60% of it) = 18; 18 x 9 = 162; 162 : 54 = 3; 3 + 180 = 183; 183 (2/3 of it) = 122;

#469. 156 + 169 = 325; 325 (80% of it) = 260; 260 (4/5 of it) = 208; 208 - 172 = 36; 36 x 6 = 216; 216 : 12 = 18;

#470. 137 + 169 = 306; 306 (50% of it) = 153; 153 - 123 = 30; 30 (4/6 of it) = 20; 20 x 10 = 200; 200 : 25 = 8;

#471. 143 : 13 = 11; 11 + 131 = 142; 142 - 117 = 25; 25 (4/5 of it) = 20; 20 x 11 = 220; 220 (30% of it) = 66;

#472. 149 + 155 = 304; 304 (50% of it) = 152; 152 (6/8 of it) = 114; 114 - 60 = 54; 54 x 11 = 594; 594 : 18 = 33;

#473. 170 : 17 = 10; 10 x 10 = 100; 100 (10% of it) = 10; 10 (square it) = 100; 100 (1/5 of it) = 20; 20 + 119 = 139;

#474. 173 + 136 = 309; 309 - 144 = 165; 165 (20% of it) = 33; 33 x 8 = 264; 264 (2/8 of it) = 66; 66 : 33 = 2;

#475. 141 (2/3 of it) = 94; 94 - 40 = 54; 54 (50% of it) = 27; 27 x 6 = 162; 162 : 81 = 2; 2 + 146 = 148;

#476. 141 : 3 = 47; 47 + 154 = 201; 201 - 151 = 50; 50 x 4 = 200; 200 (20% of it) = 40; 40 (7/8 of it) = 35;

#477. 168 : 14 = 12; 12 (square it) = 144; 144 - 82 = 62; 62 (50% of it) = 31; 31 + 174 = 205; 205 (1/5 of it) = 41;

#478. 149 + 175 = 324; 324 : 4 = 81; 81 - 26 = 55; 55 (40% of it) = 22; 22 x 4 = 88; 88 (5/8 of it) = 55;

#479. 151 + 171 = 322; 322 - 157 = 165; 165 (40% of it) = 66; 66 x 3 = 198; 198 : 6 = 33; 33 (4/6 of it) = 22;

#480. 179 + 148 = 327; 327 - 165 = 162; 162 (5/6 of it) = 135; 135 (20% of it) = 27; 27 x 11 = 297; 297 : 99 = 3;

#481. 156 : 39 = 4; 4 (square it) = 16; 16 x 5 = 80; 80 (20% of it) = 16; 16 + 125 = 141; 141 (2/3 of it) = 94;

#482. 160 (30% of it) = 48; 48 : 12 = 4; 4 x 4 = 16; 16 + 167 = 183; 183 - 155 = 28; 28 (2/8 of it) = 7;

#483. 146 - 47 = 99; 99 : 3 = 33; 33 x 12 = 396; 396 + 184 = 580; 580 (60% of it) = 348; 348 (3/4 of it) = 261;

#484. 176 : 22 = 8; 8 (square it) = 64; 64 x 7 = 448; 448 (5/8 of it) = 280; 280 (70% of it) = 196; 196 + 125 = 321;

#485. 170 (30% of it) = 51; 51 x 6 = 306; 306 (1/6 of it) = 51; 51 : 3 = 17; 17 + 125 = 142; 142 - 102 = 40;

#486. 159 + 186 = 345; 345 (80% of it) = 276; 276 : 69 = 4; 4 (square it) = 16; 16 x 4 = 64; 64 (1/4 of it) = 16;

#487. 164 - 131 = 33; 33 : 3 = 11; 11 + 189 = 200; 200 (30% of it) = 60; 60 (5/6 of it) = 50; 50 x 12 = 600;

#488. 145 (80% of it) = 116; 116 - 81 = 35; 35 (1/7 of it) = 5; 5 + 181 = 186; 186 : 31 = 6; 6 x 11 = 66;

#489. 192 (7/8 of it) = 168; 168 - 148 = 20; 20 x 9 = 180; 180 (20% of it) = 36; 36 : 18 = 2; 2 + 156 = 158;

#490. 164 - 135 = 29; 29 + 139 = 168; 168 : 6 = 28; 28 (5/7 of it) = 20; 20 x 10 = 200; 200 (10% of it) = 20;

#491. 169 - 154 = 15; 15 (square it) = 225; 225 (2/5 of it) = 90; 90 (70% of it) = 63; 63 : 7 = 9; 9 x 7 = 63;

#492. 184 (2/8 of it) = 46; 46 x 11 = 506; 506 - 161 = 345; 345 (60% of it) = 207; 207 : 69 = 3; 3 + 175 = 178;

#493. 192 - 127 = 65; 65 : 5 = 13; 13 + 149 = 162; 162 (50% of it) = 81; 81 x 4 = 324; 324 (5/6 of it) = 270;

#494. 178 - 143 = 35; 35 (80% of it) = 28; 28 x 4 = 112; 112 (3/7 of it) = 48; 48 : 12 = 4; 4 (square it) = 16;

#495. 164 (50% of it) = 82; 82 + 134 = 216; 216 : 6 = 36; 36 (2/8 of it) = 9; 9 x 9 = 81; 81 - 28 = 53;

#496. 184 (50% of it) = 92; 92 (3/4 of it) = 69; 69 : 23 = 3; 3 + 185 = 188; 188 - 173 = 15; 15 x 12 = 180;

#497. 156 - 108 = 48; 48 x 4 = 192; 192 (5/6 of it) = 160; 160 (40% of it) = 64; 64 : 32 = 2; 2 + 139 = 141;

#498. 195 : 3 = 65; 65 - 51 = 14; 14 (square it) = 196; 196 (4/7 of it) = 112; 112 + 183 = 295; 295 (80% of it) = 236;

#499. 170 (40% of it) = 68; 68 - 54 = 14; 14 x 6 = 84; 84 (3/7 of it) = 36; 36 : 3 = 12; 12 (square it) = 144;

#500. 166 + 161 = 327; 327 - 186 = 141; 141 (4/6 of it) = 94; 94 x 6 = 564; 564 : 2 = 282; 282 (50% of it) = 141;

#501. 176 (50% of it) = 88; 88 : 4 = 22; 22 x 4 = 88; 88 (1/8 of it) = 11; 11 + 167 = 178; 178 - 24 = 154;

#502. 171 - 156 = 15; 15 x 12 = 180; 180 : 12 = 15; 15 + 195 = 210; 210 (20% of it) = 42; 42 (6/7 of it) = 36;

#503. 198 (2/3 of it) = 132; 132 (50% of it) = 66; 66 - 52 = 14; 14 x 6 = 84; 84 : 12 = 7; 7 + 195 = 202;

#504. 195 (2/5 of it) = 78; 78 x 12 = 936; 936 - 164 = 772; 772 : 2 = 386; 386 (50% of it) = 193; 193 + 137 = 330;

#505. 181 + 181 = 362; 362 - 137 = 225; 225 (60% of it) = 135; 135 (2/5 of it) = 54; 54 x 9 = 486; 486 : 9 = 54;

#506. 172 : 43 = 4; 4 + 161 = 165; 165 (40% of it) = 66; 66 - 39 = 27; 27 x 9 = 243; 243 (2/3 of it) = 162;

#507. 169 : 13 = 13; 13 + 192 = 205; 205 (80% of it) = 164; 164 (6/8 of it) = 123; 123 - 51 = 72; 72 x 12 = 864;

#508. 153 (4/6 of it) = 102; 102 (50% of it) = 51; 51 x 4 = 204; 204 : 68 = 3; 3 + 175 = 178; 178 - 151 = 27;

#509. 161 : 23 = 7; 7 + 149 = 156; 156 (2/8 of it) = 39; 39 x 11 = 429; 429 - 193 = 236; 236 (50% of it) = 118;

#510. 184 (50% of it) = 92; 92 x 12 = 1104; 1104 (3/4 of it) = 828; 828 - 159 = 669; 669 : 3 = 223; 223 + 150 = 373;

#511. 195 : 13 = 15; 15 x 9 = 135; 135 (4/5 of it) = 108; 108 - 35 = 73; 73 + 142 = 215; 215 (80% of it) = 172;

#512. 187 : 17 = 11; 11 + 149 = 160; 160 (2/8 of it) = 40; 40 (10% of it) = 4; 4 (square it) = 16; 16 x 5 = 80;

#513. 180 (30% of it) = 54; 54 (5/6 of it) = 45; 45 : 3 = 15; 15 (square it) = 225; 225 - 195 = 30; 30 x 6 = 180;

#514. 184 (1/8 of it) = 23; 23 + 169 = 192; 192 - 143 = 49; 49 x 4 = 196; 196 (50% of it) = 98; 98 : 49 = 2;

#515. 165 - 101 = 64; 64 (50% of it) = 32; 32 : 8 = 4; 4 x 11 = 44; 44 (1/4 of it) = 11; 11 + 157 = 168;

#516. 200 (80% of it) = 160; 160 (6/8 of it) = 120; 120 - 51 = 69; 69 x 12 = 828; 828 : 46 = 18; 18 + 162 = 180;

#517. 202 + 204 = 406; 406 (6/7 of it) = 348; 348 - 186 = 162; 162 (50% of it) = 81; 81 : 9 = 9; 9 x 13 = 117;

#518. 177 - 154 = 23; 23 + 177 = 200; 200 (3/4 of it) = 150; 150 : 6 = 25; 25 x 6 = 150; 150 (80% of it) = 120;

#519. 165 (40% of it) = 66; 66 x 7 = 462; 462 : 22 = 21; 21 (4/7 of it) = 12; 12 + 162 = 174; 174 - 25 = 149;

#520. 180 (60% of it) = 108; 108 : 2 = 54; 54 + 184 = 238; 238 (1/7 of it) = 34; 34 x 9 = 306; 306 - 190 = 116;

#521. 183 (2/3 of it) = 122; 122 - 90 = 32; 32 (50% of it) = 16; 16 x 12 = 192; 192 : 8 = 24; 24 + 161 = 185;

#522. 174 (4/6 of it) = 116; 116 - 58 = 58; 58 x 7 = 406; 406 (50% of it) = 203; 203 : 29 = 7; 7 + 174 = 181;

#523. 208 (7/8 of it) = 182; 182 (50% of it) = 91; 91 + 199 = 290; 290 - 210 = 80; 80 x 6 = 480; 480 : 80 = 6;

#524. 182 + 150 = 332; 332 (50% of it) = 166; 166 - 54 = 112; 112 : 2 = 56; 56 (1/7 of it) = 8; 8 x 11 = 88;

#525. 206 : 2 = 103; 103 + 142 = 245; 245 (80% of it) = 196; 196 - 182 = 14; 14 x 9 = 126; 126 (6/7 of it) = 108;

#526. 206 + 179 = 385; 385 (3/7 of it) = 165; 165 (60% of it) = 99; 99 - 23 = 76; 76 : 2 = 38; 38 x 12 = 456;

#527. 212 : 4 = 53; 53 + 206 = 259; 259 - 149 = 110; 110 (20% of it) = 22; 22 x 4 = 88; 88 (2/8 of it) = 22;

#528. 193 + 212 = 405; 405 : 9 = 45; 45 (2/5 of it) = 18; 18 x 13 = 234; 234 (50% of it) = 117; 117 - 55 = 62;

#529. 182 (1/7 of it) = 26; 26 x 13 = 338; 338 (50% of it) = 169; 169 - 136 = 33; 33 : 11 = 3; 3 + 161 = 164;

#530. 192 (6/8 of it) = 144; 144 : 9 = 16; 16 x 8 = 128; 128 + 172 = 300; 300 (60% of it) = 180; 180 - 159 = 21;

#531. 174 - 87 = 87; 87 x 11 = 957; 957 (4/6 of it) = 638; 638 (50% of it) = 319; 319 : 11 = 29; 29 + 177 = 206;

#532. 181 + 206 = 387; 387 : 3 = 129; 129 - 78 = 51; 51 (2/3 of it) = 34; 34 x 5 = 170; 170 (50% of it) = 85;

#533. 184 - 148 = 36; 36 x 8 = 288; 288 : 8 = 36; 36 (1/4 of it) = 9; 9 + 196 = 205; 205 (40% of it) = 82;

#534. 203 (2/7 of it) = 58; 58 x 13 = 754; 754 (50% of it) = 377; 377 - 208 = 169; 169 : 13 = 13; 13 + 189 = 202;

#535. 197 + 211 = 408; 408 (2/3 of it) = 272; 272 - 206 = 66; 66 x 13 = 858; 858 (50% of it) = 429; 429 : 39 = 11;

#536. 215 (2/5 of it) = 86; 86 + 169 = 255; 255 (40% of it) = 102; 102 : 2 = 51; 51 x 9 = 459; 459 - 211 = 248;

#537. 201 - 157 = 44; 44 : 11 = 4; 4 + 184 = 188; 188 (50% of it) = 94; 94 x 4 = 376; 376 (7/8 of it) = 329;

#538. 183 (2/3 of it) = 122; 122 (50% of it) = 61; 61 + 187 = 248; 248 : 62 = 4; 4 (square it) = 16; 16 x 9 = 144;

#539. 196 : 14 = 14; 14 x 5 = 70; 70 + 215 = 285; 285 (3/5 of it) = 171; 171 - 29 = 142; 142 (50% of it) = 71;

#540. 212 : 2 = 106; 106 - 71 = 35; 35 x 13 = 455; 455 (4/7 of it) = 260; 260 (40% of it) = 104; 104 + 174 = 278;

#541. 218 - 193 = 25; 25 (2/5 of it) = 10; 10 (square it) = 100; 100 : 4 = 25; 25 (80% of it) = 20; 20 x 7 = 140;

#542. 175 (5/7 of it) = 125; 125 : 25 = 5; 5 + 179 = 184; 184 (50% of it) = 92; 92 x 4 = 368; 368 - 157 = 211;

#543. 214 - 181 = 33; 33 x 11 = 363; 363 (4/6 of it) = 242; 242 : 22 = 11; 11 + 159 = 170, 170 (80% of it) = 136;

#544. 178 : 89 = 2; 2 + 151 = 153; 153 - 123 = 30; 30 x 12 = 360; 360 (2/3 of it) = 240; 240 (20% of it) = 48;

#545. 196 (2/8 of it) = 49; 49 x 13 = 637; 637 - 219 = 418; 418 (50% of it) = 209; 209 : 11 = 19; 19 + 184 = 203;

#546. 186 (50% of it) = 93; 93 - 61 = 32; 32 : 2 = 16; 16 x 4 = 64; 64 (1/4 of it) = 16; 16 + 181 = 197;

#547. 181 + 214 = 395; 395 - 199 = 196; 196 : 14 = 14; 14 (square it) = 196; 196 (5/7 of it) = 140; 140 (20% of it) = 28;

#548. 189 (2/7 of it) = 54; 54 : 2 = 27; 27 + 203 = 230; 230 (20% of it) = 46; 46 x 3 = 138; 138 - 98 = 40;

#549. 215 (40% of it) = 86; 86 x 9 = 774; 774 (1/6 of it) = 129; 129 + 167 = 296; 296 : 37 = 8; 8 (square it) = 64;

#550. 194 : 97 = 2; 2 + 212 = 214; 214 - 174 = 40; 40 (70% of it) = 28; 28 x 4 = 112; 112 (6/8 of it) = 84;

#551. 217 - 197 = 20; 20 x 8 = 160; 160 (70% of it) = 112; 112 (6/7 of it) = 96; 96 : 48 = 2; 2 + 168 = 170;

#552. 185 (60% of it) = 111; 111 + 173 = 284; 284 (3/4 of it) = 213; 213 - 39 = 174; 174 : 29 = 6; 6 (square it) = 36;

#553. 200 (3/8 of it) = 75; 75 (80% of it) = 60; 60 : 10 = 6; 6 x 5 = 30; 30 + 173 = 203; 203 - 168 = 35;

#554. 198 : 66 = 3; 3 + 163 = 166; 166 - 43 = 123; 123 (2/3 of it) = 82; 82 x 5 = 410; 410 (30% of it) = 123;

#555. 208 (3/8 of it) = 78; 78 - 32 = 46; 46 x 10 = 460; 460 (50% of it) = 230; 230 : 10 = 23; 23 + 197 = 220;

#556. 201 - 162 = 39; 39 (4/6 of it) = 26; 26 x 3 = 78; 78 : 3 = 26; 26 (50% of it) = 13; 13 + 191 = 204;

#557. 193 + 212 = 405; 405 - 162 = 243; 243 : 3 = 81; 81 x 10 = 810; 810 (10% of it) = 81; 81 (4/6 of it) = 54;

#558. 210 (70% of it) = 147; 147 : 7 = 21; 21 (2/3 of it) = 14; 14 x 7 = 98; 98 - 40 = 58; 58 + 200 = 258;

#559. 185 : 5 = 37; 37 + 209 = 246; 246 (2/3 of it) = 164; 164 (50% of it) = 82; 82 x 10 = 820; 820 - 166 = 654;

#560. 218 : 2 = 109; 109 + 216 = 325; 325 - 165 = 160; 160 (30% of it) = 48; 48 (5/8 of it) = 30; 30 x 3 = 90;

#561. 217 - 203 = 14; 14 x 5 = 70; 70 (20% of it) = 14; 14 (square it) = 196; 196 (2/7 of it) = 56; 56 + 184 = 240;

#562. 218 : 2 = 109; 109 + 216 = 325; 325 (1/5 of it) = 65; 65 x 8 = 520; 520 (50% of it) = 260; 260 - 184 = 76;

#563. 190 (30% of it) = 57; 57 (2/3 of it) = 38; 38 x 5 = 190; 190 + 175 = 365; 365 - 175 = 190; 190 : 19 = 10;

#564. 181 + 199 = 380; 380 : 4 = 95; 95 x 8 = 760; 760 (70% of it) = 532; 532 (6/8 of it) = 399; 399 - 161 = 238;

#565. 212 : 53 = 4; 4 + 162 = 166; 166 - 115 = 51; 51 x 12 = 612; 612 (50% of it) = 306; 306 (5/6 of it) = 255;

#566. 228 (2/3 of it) = 152; 152 (50% of it) = 76; 76 : 19 = 4; 4 (square it) = 16; 16 + 164 = 180; 180 - 90 = 90;

#567. 209 : 11 = 19; 19 + 169 = 188; 188 (50% of it) = 94; 94 - 79 = 15; 15 x 13 = 195; 195 (2/3 of it) = 130;

#568. 199 + 185 = 384; 384 (50% of it) = 192; 192 : 32 = 6; 6 (square it) = 36; 36 (4/6 of it) = 24; 24 x 6 = 144;

#569. 232 (5/8 of it) = 145; 145 (60% of it) = 87; 87 x 13 = 1131; 1131 : 3 = 377; 377 - 180 = 197; 197 + 194 = 391;

#570. 230 (10% of it) = 23; 23 + 220 = 243; 243 (4/6 of it) = 162; 162 - 141 = 21; 21 x 6 = 126; 126 : 21 = 6;

#571. 192 (50% of it) = 96; 96 (3/4 of it) = 72; 72 : 18 = 4; 4 (square it) = 16; 16 + 166 = 182; 182 - 150 = 32;

#572. 215 - 167 = 48; 48 (50% of it) = 24; 24 (3/8 of it) = 9; 9 x 11 = 99; 99 : 9 = 11; 11 + 228 = 239;

#573. 195 - 181 = 14; 14 x 13 = 182; 182 (4/7 of it) = 104; 104 (50% of it) = 52; 52 : 26 = 2; 2 + 203 = 205;

#574. 233 + 185 = 418; 418 - 202 = 216; 216 (5/6 of it) = 180; 180 (30% of it) = 54; 54 : 3 = 18; 18 x 4 = 72;

#575. 193 + 172 = 365; 365 (3/5 of it) = 219; 219 - 43 = 176; 176 : 22 = 8; 8 (square it) = 64; 64 (50% of it) = 32;

#576. 209 : 11 = 19; 19 + 233 = 252; 252 - 188 = 64; 64 (50% of it) = 32; 32 (1/8 of it) = 4; 4 x 12 = 48;

#577. 203 (2/7 of it) = 58; 58 : 29 = 2; 2 + 200 = 202; 202 - 42 = 160; 160 (20% of it) = 32; 32 x 6 = 192;

#578. 202 : 2 = 101; 101 + 219 = 320; 320 (1/4 of it) = 80; 80 (70% of it) = 56; 56 x 8 = 448; 448 - 229 = 219;

#579. 192 : 2 = 96; 96 (1/8 of it) = 12; 12 x 14 = 168; 168 + 172 = 340; 340 (50% of it) = 170; 170 - 100 = 70;

#580. 208 (2/8 of it) = 52; 52 (50% of it) = 26; 26 x 4 = 104; 104 + 204 = 308; 308 - 209 = 99; 99 : 3 = 33;

#581. 202 + 221 = 423; 423 - 228 = 195; 195 (60% of it) = 117; 117 : 3 = 39; 39 x 10 = 390; 390 (2/3 of it) = 260;

#582. 198 - 184 = 14; 14 (square it) = 196; 196 (1/4 of it) = 49; 49 : 7 = 7; 7 + 218 = 225; 225 (40% of it) = 90;

#583. 222 : 37 = 6; 6 x 4 = 24; 24 (1/6 of it) = 4; 4 + 214 = 218; 218 - 193 = 25; 25 (60% of it) = 15;

#584. 234 (4/6 of it) = 156; 156 (50% of it) = 78; 78 : 26 = 3; 3 + 234 = 237; 237 - 182 = 55; 55 x 9 = 495;

#585. 223 + 188 = 411; 411 (2/3 of it) = 274; 274 - 202 = 72; 72 : 6 = 12; 12 x 3 = 36; 36 (50% of it) = 18;

#586. 222 : 74 = 3; 3 + 204 = 207; 207 - 169 = 38; 38 x 5 = 190; 190 (80% of it) = 152; 152 (1/8 of it) = 19;

#587. 202 (50% of it) = 101; 101 + 225 = 326; 326 - 214 = 112; 112 : 2 = 56; 56 x 8 = 448; 448 (1/8 of it) = 56;

#588. 209 : 11 = 19; 19 + 215 = 234; 234 (2/3 of it) = 156; 156 - 93 = 63; 63 x 8 = 504; 504 (50% of it) = 252;

#589. 213 + 205 = 418; 418 : 11 = 38; 38 x 3 = 114; 114 (5/6 of it) = 95; 95 (60% of it) = 57; 57 - 28 = 29;

#590. 216 (7/8 of it) = 189; 189 - 64 = 125; 125 (20% of it) = 25; 25 x 11 = 275; 275 : 25 = 11; 11 + 240 = 251;

#591. 230 - 184 = 46; 46 x 7 = 322; 322 (3/7 of it) = 138; 138 (50% of it) = 69; 69 : 23 = 3; 3 + 226 = 229;

#592. 195 - 175 = 20; 20 x 12 = 240; 240 (50% of it) = 120; 120 (4/6 of it) = 80; 80 : 5 = 16; 16 + 199 = 215;

#593. 232 - 194 = 38; 38 (50% of it) = 19; 19 + 197 = 216; 216 (3/8 of it) = 81; 81 : 9 = 9; 9 x 4 = 36;

#594. 208 (1/8 of it) = 26; 26 x 13 = 338; 338 : 2 = 169; 169 + 205 = 374; 374 (50% of it) = 187; 187 - 104 = 83;

#595. 226 (50% of it) = 113; 113 + 177 = 290; 290 (4/5 of it) = 232; 232 : 4 = 58; 58 x 5 = 290; 290 - 185 = 105;

#596. 236 (50% of it) = 118; 118 - 74 = 44; 44 (3/4 of it) = 33; 33 x 10 = 330; 330 : 22 = 15; 15 + 179 = 194;

#597. 247 + 237 = 484; 484 (50% of it) = 242; 242 : 2 = 121; 121 - 61 = 60; 60 x 14 = 840; 840 (3/4 of it) = 630;

#598. 222 : 2 = 111; 111 - 80 = 31; 31 + 189 = 220; 220 (2/8 of it) = 55; 55 x 12 = 660; 660 (80% of it) = 528;

#599. 218 (50% of it) = 109; 109 + 186 = 295; 295 (3/5 of it) = 177; 177 - 81 = 96; 96 : 16 = 6; 6 x 7 = 42;

#600. 204 (4/6 of it) = 136; 136 + 206 = 342; 342 (50% of it) = 171; 171 : 3 = 57; 57 - 42 = 15; 15 x 5 = 75;

Printed in Great Britain
by Amazon

32237345R00059